Low-Visibility Antennas for Communication Systems

GREGORY L. CHARVAT SERIES ON PRACTICAL APPROACHES TO ELECTRICAL ENGINEERING

SERIES EDITOR

Gregory L. Charvat
Advisor, Co-Founder, Researcher,
Westbrook, Connecticut, USA

PUBLISHED TITLES

Low-Visibility Antennas for Communication Systems
Albert Sabban

Small and Short-Range Radar Systems
Gregory L. Charvat

Low-Visibility Antennas for Communication Systems

DR. ALBERT SABBAN

Braude College, Karmiel, Israel

CRC Press
Taylor & Francis Group
Boca Raton London New York

CRC Press is an imprint of the
Taylor & Francis Group, an **informa** business

CRC Press
Taylor & Francis Group
6000 Broken Sound Parkway NW, Suite 300
Boca Raton, FL 33487-2742

First issued in paperback 2017

ISBN-13: 978-1-4822-4643-8 (hbk)
ISBN-13: 978-1-138-74810-1 (pbk)

Library of Congress Cataloging-in-Publication Data

Sabban, Albert.
    Low-visibility antennas for communication systems / author, Albert Sabban.
        pages cm -- (Gregory L. Charvat series on practical approaches to electrical
        engineering)
    "A CRC title."
    Includes bibliographical references and index.
    ISBN 978-1-4822-4643-8
    1. Microstrip antennas. 2. Wireless communication systems--Equipment and
supplies. 3. Miniature electronic equipment. I. Title.

TK7871.67.M5S235 2015
621.382'4--dc23
                                                    2015012296

Visit the Taylor & Francis Web site at
http://www.taylorandfrancis.com

and the CRC Press Web site at
http://www.crcpress.com

The book is dedicated to the memory of my father, mother, and sister:
David Sabban, Dolly Sabban, and Aliza Sabban
Also dedicated to my family:
My wife, Mazal Sabban
My daughters, Dolly and Lilach
My son, David Sabban
Grandchildren, Nooa, Avigail, Ido, and Shirra

Contents

Series Preface

GREGORY L. CHARVAT SERIES ON PRACTICAL APPROACHES TO ELECTRICAL ENGINEERING

Learn, create, execute. Technology development moves quickly in the modern era. An engineer must "ballpark" a design, rapidly create proof of concept, and then measure results to drive product development. The purpose of this series is to enable those interested in moving from concept to practical implementation. Key concepts and background will be covered, with tangential discussion minimized. This series of texts will serve to bridge the gap between academia and implementation with an emphasis on practical approaches.

Preface

The main objective of *Low-Visibility Antennas for Communication Systems* is to present new low-visibility antennas. To provide a background, Chapters 1 and 2 cover electromagnetics and basic theory and fundamentals of antennas to assist electrical engineers and students. There are many electromagnetic theory and antenna books for electromagnetic scientists. However, there are few books that help electrical engineers and undergraduate students to study and understand basic electromagnetic and antennas theory and fundamentals with few integral and differential equations.

Several types of 3D full-wave electromagnetics software are available, such as the High-Frequency Structural Simulator (originally, now HFSS), Advanced Design System (ADS), and Computer Simulation Technology (CST) used to design and analyze antennas. Antennas developed and analyzed in this book were designed by using HFSS and ADS software. For all the antennas presented in this book there was a good agreement between computed and measured results. Only one design and fabrication iteration was needed in the development process of the antenna presented.

Electromagnetic theory and transmission lines are discussed in Chapter 1. An introduction to antennas is given in Chapter 2. Low-visibility printed antennas are presented in Chapter 3. Antenna arrays are the subject of Chapter 4. Applications of microstrip antenna arrays are discussed in Chapter 5. Also, millimeter wave (MM) microstrip antennas are presented in Chapter 5. Wearable antennas for medical applications are presented in Chapter 6, and antenna S_{11} variation as a function of distance from the human body is discussed in Section 6.4. Wearable tunable printed antennas for medical applications are presented in Chapter 7. New wideband wearable meta-material antennas for communication applications are presented in Chapter 8, and meta-material antenna characteristics in the vicinity of the human body are discussed in Section 8.6. Low-visibility fractal printed antennas are presented in Chapter 9. Microwave and MM wave technologies—microwave integrated circuits (MIC), monolithic microwave integrated circuits (MMIC), micro-electro-mechanical systems (MEMS), and low-temperature co-fired ceramics (LTCC)—are presented in Chapter 10. Radio frequency and antenna measurements are presented in Chapter 11.

Each chapter in the book covers sufficient details to enable students, scientists from all areas, and electrical and biomedical engineers to follow and understand the topics covered in the book. The book begins with elementary electromagnetics and antenna topics needed by students and engineers with no background in electromagnetic and antenna theory to study and understand the basic design principles and features of antennas, printed antennas, compact antennas with low visibility, and wearable antennas for communication and medical applications. This book may serve students and design engineers as a reference book.

Several topics and designs are presented in this book for the first time. These include new designs in the areas of wearable antennas, meta-material antennas, and fractal antennas. The text contains sufficient mathematical detail and explanations to enable electrical engineering and physics students to understand all topics presented in this book.

Several new antennas are presented in this book. Design considerations and computed and measured results of the new antennas are presented in the book.

About the Author

Dr. Albert Sabban earned the BSc degree and MSc degree (magna cum laude) in electrical engineering from Tel Aviv University, Israel in 1976 and 1986, respectively. His MSc thesis was "Spectral Domain Iterative Analysis of Multilayered Microstrip Antennas." He earned the PhD degree in electrical engineering from Colorado University at Boulder, Colorado, in 1991. His PhD dissertation was "Multiport Network Model for Evaluating Radiation Loss and Coupling among Discontinuities in Microstrip Circuits." Dr. Sabban's research interests include microwave and antenna engineering, biomedical engineering, and communication and system engineering.

Dr. Sabban is a senior lecturer, Department of Electrical and Electronic Engineering, Ort Braude Engineering College, Karmiel, Israel. From 1976 to 2007, Dr. Sabban was a senior R&D scientist and project leader in RAFAEL. He successfully passed a system engineering course in RAFAEL. During his work in RAFAEL and other institutes and companies Dr. Sabban gained experience in project management, sales, marketing, and training. He managed and led groups and projects with more than 20 employees.

Dr. Sabban developed radio frequency integrated circuit (RFIC) components on gallium arsenide (GaAs) and silicon substrates. He developed microwave components by employing LTCC technology. In RAFAEL he developed passive and active microwave components such as power amplifiers, low-noise amplifiers, multipliers, voltage-controlled oscillators (VCOs), power dividers filters, and RF heads. Dr. Sabban developed 20 W, 100 W power amplifiers at ultra-high frequencies. He developed a 10 W class C transmitter at L band frequencies with 50% efficiency. He developed a 2 W transmitter at Ka band. He also developed wideband microstrip antenna arrays, dipole antenna arrays, telemetry microstrip antennas, back-fire antennas, reflector antennas, couplers, power dividers, and wideband monopulse comparators. From 1979 to 1984 Dr. Sabban was a teaching assistant in the Electrical Engineering Department in the Technion in Haifa in Israel. From 1984 to 1987 he was the leader of an antenna R&D group in RAFAEL. From 1980 to 1987 he was a leader of several research programs in RAFAEL. In his research programs he developed wideband MM wave microstrip antennas, compact reflector antennas at MM wave frequencies, dielectric rod antennas, and passive microwave components. He also developed an iterative spectral domain analysis of single- and double-layered microstrip antennas using the conjugate gradient algorithm. From August 1987 to February 1991 he was on leave from RAFAEL and joined the University of Colorado at Boulder, where he studied toward his PhD degree. He worked as a research assistant in the microwave and millimeter wave computer-aided design center in the University of Colorado at Boulder. His research topic was "Multiport Network Modeling for Evaluating Radiation Loss and Spurious Coupling among Microstrip Discontinuities in Microstrip Circuits." Dr. Sabban developed a planar lumped model for evaluating spurious coupling and radiation loss among coupled microstrip discontinuities, a spectral domain algorithm to analyze microstrip lines and coupled microstrip lines, and an iterative spectral domain algorithm to analyze wideband microstrip antennas. He holds a U.S. patent on wideband microstrip antennas. Since March 1991, Dr. Sabban has been working as the leading R&D microwave scientist at RAFAEL, where he developed low-noise amplifiers, power amplifiers, and passive microwave components. From 1991 to August 1993 he worked as a leading R&D engineer in developing RF environmental simulation systems. From August 1993 to August 1994 he worked as a leading R&D engineer in developing a compact low-power consumption integrated RF head for the Inmarsat-M Ground Terminal. From August 1995 to August 1998 he worked as a senior project leader in developing a compact low-power consumption integrated RF head at Ka band for very small aperture terminal (VSAT) applications. From 1988 to July 2000 Dr. Sabban worked as a senior R&D scientist and project leader in developing MMIC RF components and RF heads at Ka band. From July 2000 to 2001 he developed RFIC components on GaAs and silicon substrates. From June 2001 to 2002 he developed microwave components by employing LTCC technology. Since 2002 Dr. Sabban has been working as a senior R&D scientist and project leader in developing MMIC RF modules and RF systems at Ka band. At the beginning of 2005 Dr. Sabban joined the MEMS group in RAFAEL and led and developed projects in the MEMS group.

He published over 60 research papers in journals and conferences and hold a patent in the antenna area.

Dr. Sabban wrote chapters in books on printed antennas. He published in Israel a book on electromagnetic and transmission line theory. His papers are cited in several books and papers.

1 Electromagnetic Theory and Transmission Lines

1.1 DEFINITIONS

Angular frequency: The angular frequency, ω, represents the frequency in radians per second. $\omega = v*k = 2\pi f$.

Antenna: An antenna is used to radiate efficiently electromagnetic energy in desired directions. Antennas match radio frequency systems to space. All antennas may be used to receive or radiate energy. Antennas transmit or receive electromagnetic waves. Antennas convert electromagnetic radiation into electric current, or vice versa. Antennas transmit and receive electromagnetic radiation at radio frequencies. Antennas are a necessary part of all communication links and radio equipment. Antennas are used in systems such as radio and television broadcasting, point-to-point radio communication, wireless local area network (LAN), cell phones, radar, medical systems, and spacecraft communication. Antennas are most commonly employed in air or outer space, but can also be operated under water, on and inside the human body, or even through soil and rock at low frequencies for short distances.

Field: A field is a physical quantity that has a value for each point in space and time.

Frequency: The frequency, f, is the number of periods per unit time (second) and is measured in hertz.

Phase velocity: The phase velocity, v, of a wave is the rate at which the phase of the wave propagates in space. Phase velocity is measured in m/s.

$$\lambda = v * T$$
$$T = 1/f \tag{1.1}$$
$$\lambda = v/f$$

Electromagnetic waves propagate in free space in the phase velocity of light.

$$v = 3 \cdot 10^8 \text{ m/s} \tag{1.2}$$

Polarization: A wave is polarized if it oscillates in one direction or plane. The polarization of a transverse wave describes the direction of oscillation in the plane perpendicular to the direction of propagation.

1

Wavelength: The wavelength is the distance between two sequential equivalent points. Wavelength, λ, is measured in meters.

Wavenumber: A wavenumber, k, is the spatial frequency of the wave in radians per unit distance (per meter). $k = 2\pi/\lambda$.

Wave period: The wave period is the time, T, for one complete cycle of an oscillation of a wave. The wave period is measured in seconds.

1.2   ELECTROMAGNETIC WAVES

1.2.1   Maxwell's Equations

Maxwell's equations describe how electric and magnetic fields are generated and altered by each other [1–8]. Maxwell's equations are a classical approximation to the more accurate and fundamental theory of quantum electrodynamics. Quantum deviations from Maxwell's equations are usually small. Inaccuracies occur when the particle nature of light is important or when electric fields are strong.

TABLE 1.1
Symbols and Abbreviations

Dimensions	Parameter	Symbol
Wb/m	Magnetic potential	A
m/s^2	Acceleration	a
Tesla	Magnetic field	B
F	Capacitance	C
V/m	Electric field displacement	D
V/m	Electric field	E
N	Force	F
A/m	Magnetic field strength	H
A	Current	I
A/m^2	Current density	J
H	Self-inductance	L
M	Length	l
H	Mutual-inductance	M
C	Charge	q
W/m^2	Poynting vector	**P**
W	Power	P
Ω	Resistance	R
m^3	Volume	V
m/s	Velocity	v
F/m	Dielectric constant	ε
	Relative dielectric constant	ε_r
H/m	Permeability	μ
$1/\Omega \cdot m$	Conductivity	σ
Wb	Magnetic flux	ψ

1.2.2 GAUSS'S LAW FOR ELECTRIC FIELDS

Gauss's law for electric fields states that the electric flux via any closed surface S is equal to the net charge q divided by the free space dielectric constant.

$$\oint_S E \cdot ds = \frac{1}{\varepsilon_0} \int_V \rho \, dv = \frac{1}{\varepsilon_0} q$$

$$D = \varepsilon E \tag{1.3}$$

$$\nabla \cdot D = \rho_v$$

1.2.3 GAUSS'S LAW FOR MAGNETIC FIELDS

Gauss's law for magnetic fields states that the magnetic flux via any closed surface S is equal to zero. There is no magnetic charge in nature.

$$\oint_S B \cdot ds = 0$$

$$B = \mu H \tag{1.4}$$

$$\psi_m = \int_S B \cdot ds$$

$$\nabla \cdot B = 0$$

1.2.4 AMPÈRE'S LAW

The original Ampère's law stated that magnetic fields can be generated by an electrical current. Ampère's law was corrected by Maxwell, who stated that magnetic fields can be generated also by time-variant electric fields. The corrected Ampère's law shows that a changing magnetic field induces an electric field and also a time-variant electric field induces a magnetic field.

$$\oint_C \frac{B}{\mu_0} \cdot dl = \int_S J \cdot ds + \frac{d}{dt} \int_s \varepsilon_0 E \cdot ds$$

$$\nabla X H = J + \frac{\partial D}{\partial t} \tag{1.5}$$

$$J = \sigma E$$

$$i = \int_S J \cdot ds$$

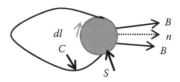

FIGURE 1.1 Faraday's law.

1.2.5 FARADAY'S LAW

Faraday's law describes how a propagating time-varying magnetic field through a surface S creates an electric field, as shown in Figure 1.1.

$$\oint_C E \cdot dl = -\frac{d}{dt}\int_S B \cdot ds$$

$$\nabla XE = -\frac{\partial B}{\partial t} \tag{1.6}$$

1.2.6 WAVE EQUATIONS

The variation of electromagnetic waves as a function of time may be written as $e^{j\omega t}$. The derivative as a function of time is $j\omega e^{j\omega t}$. Maxwell's equations may be written as

$$\nabla XE = -j\omega\mu H$$

$$\nabla XH = (\sigma + j\omega\varepsilon)E \tag{1.7}$$

A ∇X(curl) operation on the electric field E results in

$$\nabla X\nabla XE = -j\omega\mu\nabla XH \tag{1.8}$$

By substituting the expression of ∇XH in Equation 1.8 we get

$$\nabla X\nabla XE = -j\omega\mu(\sigma + j\omega\varepsilon)E \tag{1.9}$$

$$\nabla X\nabla XE = -\nabla^2 E + \nabla \left(\nabla \cdot E \right) \tag{1.10}$$

In free space there is no charge so $\nabla \cdot E = 0$. We get the wave equation for an electric field.

$$\nabla^2 E = j\omega\mu(\sigma + j\omega\varepsilon)E = \gamma^2 E$$

$$\gamma = \sqrt{j\omega\mu(\sigma + j\omega\varepsilon)} = \alpha + j\beta \tag{1.11}$$

TABLE 1.2

Symbols and Abbreviations

Parameter	Symbol	Dimensions
Electric field displacement	D	V/m
Electric field	E	V/m
Magnetic field strength	H	A/m
Dielectric constant in space	ε_0	$8.854 \cdot 10^{-12}$ F/m
Permeability in space	μ_0	$\mu_0 = 4\pi \cdot 10^{-7}$ H/m
Volume charge density	ρ_V	c/m^3
Magnetic field	B	Tesla, Wb/m^2
Conductivity	σ	$1/\Omega \cdot m$
Magnetic flux	Ψ	Wb
Skin depth	δ_s	M

where γ is the complex propagation constant; α represents losses in the medium; and β represents the wave phase constant in radians per meter. If we follow the same procedure on the magnetic field we will get the wave equation for a magnetic field.

$$\nabla^2 H = j\omega\mu(\sigma + j\omega\varepsilon)H = \gamma^2 H \tag{1.12}$$

The law of conservation of energy imposes boundary conditions on the electric and magnetic fields. When an electromagnetic wave travels from medium 1 to medium 2 the electric and magnetic fields should be continuous, as shown in Figure 1.2.

General boundary conditions

$$n \cdot (D_2 - D_1) = \rho_S \tag{1.13}$$

$$n \cdot (B_2 - B_1) = 0 \tag{1.14}$$

$$(E_2 - E_1) \times n = M_S \tag{1.15}$$

$$n \times (H_2 - H_1) = J_S \tag{1.16}$$

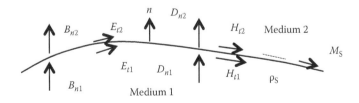

FIGURE 1.2 Fields between two media.

Boundary conditions for a dielectric medium

$$n \cdot (D_2 - D_1) = 0 \tag{1.17}$$

$$n \cdot (B_2 - B_1) = 0 \tag{1.18}$$

$$(E_2 - E_1) \times n = 0 \tag{1.19}$$

$$n \times (H_2 - H_1) = 0 \tag{1.20}$$

Boundary conditions for a conductor

$$n \cdot (D_2 - D_1) = \rho_S \tag{1.21}$$

$$n \cdot (B_2 - B_1) = 0 \tag{1.22}$$

$$(E_2 - E_1) \times n = 0 \tag{1.23}$$

$$n \times (H_2 - H_1) = J_S \tag{1.24}$$

The condition for a good conductor is $\sigma \gg \omega\varepsilon$.

$$\gamma = \sqrt{j\omega\mu(\sigma + j\omega\varepsilon)} = \alpha + j\beta \approx (1+j)\sqrt{\frac{\omega\mu\sigma}{2}} \tag{1.25}$$

$$\delta_s = \sqrt{\frac{2}{\omega\mu\sigma}} = \frac{1}{\alpha} \tag{1.26}$$

The conductor skin depth is given as δ_s. The wave attenuation is α.

1.3 TRANSMISSION LINES

Transmission lines are used to transfer electromagnetic energy from one point to another with minimum losses over a wide band of frequencies [1–8]. There are three major types of transmission lines: transmission lines with a cross section that is very small compared to the wavelength, in which the dominant mode of propagation is the transverse electromagnetic mode (TEM), and closed rectangular and cylindrical conducting tubes in which the dominant modes of propagation are the transverse electric (TE) mode and the transverse magnetic (TM) mode. Open boundary structures that their cross section is greater than 0.1 λ may support surface wave mode of propagation.

For TEM modes $E_z = H_z = 0$. For TE modes $E_z = 0$. For TM modes $H_z = 0$.

Voltage and currents in transmission lines may be derived by using the transmission lines equations. Transmission line equations may be derived by employing Maxwell equations and the boundary conditions on the transmission line section

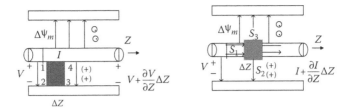

FIGURE 1.3 Transmission line geometry.

shown in Figure 1.3. Equation 1.27 is the first lossless transmission lines equation. l_e is the self-inductance per length.

$$\frac{\partial V}{\partial Z} = -\frac{\partial}{\partial t} l_e I = -l_e \frac{\partial I}{\partial t} \tag{1.27}$$

$$\frac{\partial I}{\partial Z} = -c \frac{\partial V}{\partial t} - gV \tag{1.28}$$

Equation 1.28 is the second lossless transmission lines equation.

$$c = \frac{\Delta C}{\Delta Z} F/m.$$

$$g = \frac{\Delta G}{\Delta Z} m/\Omega.$$

Equation 1.29 is the first transmission line equation with losses.

$$-\frac{\partial v}{\partial z} = Ri + L \frac{\partial i}{\partial t} \tag{1.29}$$

Equation 1.30 is the second transmission line equation with losses.

$$-\frac{\partial i}{\partial z} = Gv + C \frac{\partial v}{\partial t} \tag{1.30}$$

where

$$C = \frac{\Delta C(z)}{\Delta Z} F/m \quad L = \frac{\Delta L(z)}{\Delta Z} H/m \quad R = \frac{1}{G} = \frac{\Delta R(z)}{\Delta Z} m/\Omega$$

By differentiating Equation 1.29 with respect to z and by differentiating Equation 1.30 with respect to t and adding the result we get

$$-\frac{\partial^2 v}{\partial z^2} = RGv + (RC + LG)\frac{\partial v}{\partial t} + LC\frac{\partial^2 v}{\partial t^2} \qquad (1.31)$$

By differentiating Equation 1.30 with respect to z and by differentiating Equation 1.29 with respect to t and adding the result we get

$$-\frac{\partial^2 i}{\partial z^2} = RGi + (RC + LG)\frac{\partial i}{\partial t} + LC\frac{\partial^2 i}{\partial t^2} \qquad (1.32)$$

Equations 1.31 and 1.32 are analog to the wave equations. The solution of these equations is a superposition of a forward wave, $+z$, and backward wave, $-z$.

$$V(z,t) = V_+\left(t - \frac{z}{v}\right) + V_-\left(t + \frac{z}{v}\right)$$

$$I(z,t) = Y_0\left\{V_+\left(t - \frac{z}{v}\right) - V_-\left(t + \frac{z}{v}\right)\right\} \qquad (1.33)$$

where Y_0 is the characteristic admittance of the transmission line $Y_0 = \dfrac{1}{Z_0}$. The variation of electromagnetic waves as function of time may be written as $e^{j\omega t}$. The derivative as function of time is $j\omega e^{j\omega t}$. By using these relations we may write phazor transmission line equations.

$$\frac{dV}{dZ} = -ZI$$

$$\frac{dI}{dZ} = -YV$$

$$\frac{d^2V}{dz^2} = \gamma^2 V \qquad (1.34)$$

$$\frac{d^2I}{dz^2} = \gamma^2 I$$

where

$$Z = R + j\omega L \quad \Omega/m$$

$$Y = G + j\omega C \quad m/\Omega \qquad (1.35)$$

$$\gamma = \alpha + j\beta = \sqrt{ZY}$$

The solution of the transmission lines equations in harmonic steady state is

$$v(z,t) = \operatorname{Re} V(z)e^{j\omega t}$$
$$i(z,t) = \operatorname{Re} I(z)e^{j\omega t}$$

(1.36)

$$V(z) = V_+ e^{-\gamma z} + V_- e^{\gamma z}$$
$$I(z) = I_+ e^{-\gamma z} + I_- e^{\gamma z}$$

(1.37)

For a lossless transmission line we may write

$$\frac{dV}{dZ} = -j\omega L I$$

$$\frac{dI}{dZ} = -j\omega C V$$

$$\frac{d^2 V}{dz^2} = -\omega^2 L C V$$

$$\frac{d^2 I}{dz^2} = -\omega^2 L C I$$

(1.38)

The solution of the lossless transmission lines equations is

$$V(z) = e^{j\omega t}\left(V^+ e^{-j\beta z} + V^- e^{j\beta z}\right)$$
$$I(z) = Y_0\left(e^{j\omega t}\left(V^+ e^{-j\beta z} - V^- e^{j\beta z}\right)\right)$$

(1.39)

$$v_p = \frac{\omega}{\beta} = \frac{1}{\sqrt{LC}} = \frac{1}{\sqrt{\mu\varepsilon}}$$

where
v_p = the phase velocity
Z_0 = the characteristic impedance of the transmission line.

$$Z_0 = \frac{V_+}{I_+} = \frac{V_-}{I_-} = \sqrt{\frac{(R + j\omega L)}{(G + j\omega C)}}$$

for $R = 0$ $G = 0$

(1.40)

$$Z_0 = \frac{V_+}{I_+} = \frac{V_-}{I_-} = \sqrt{\frac{L}{C}}$$

Waves in transmission lines

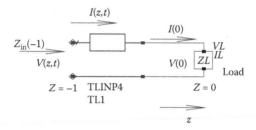

FIGURE 1.4 Transmission line with load.

A load Z_L is connected, at $z = 0$, to a transmission line with impedance Z_0 (Figure 1.4). The voltage on the load is V_L. The current on the load is I_L.

$$V(0) = V_L = I(0) \cdot Z_L$$
$$I(0) = I_L \tag{1.41}$$

For $z = 0$ we can write

$$V(0) = I(0) \cdot Z_L = V_+ + V_-$$
$$I(0) = Y_0 (V_+ - V_-) \tag{1.42}$$

By substituting $I(0)$ in $V(0)$ we get

$$Z_L = Z_0 \frac{V_+ + V_-}{(V_+ - V_-)}$$

$$Z_L = Z_0 \frac{1 + \dfrac{V_-}{V_+}}{1 - \dfrac{V_-}{V_+}} \tag{1.43}$$

The ratio $\dfrac{V_-}{V_+}$ is defined as reflection coefficient.

$$\Gamma_L = \frac{V_-}{V_+}.$$

$$Z_L = Z_0 \frac{1 + \Gamma_L}{1 - \Gamma_L} \tag{1.44}$$

$$\Gamma_L = \frac{\dfrac{Z_L}{Z_0} - 1}{\dfrac{Z_L}{Z_0} + 1} \tag{1.45}$$

The reflection coefficient as function of z may be written as

$$\Gamma(Z) = \frac{V_-}{V_+} = \frac{V_- e^{j\beta z}}{V_+ e^{-j\beta z}} = \Gamma_L e^{2j\beta z} \tag{1.46}$$

The input impedance as function of z may be written as

$$Z_{in}(z) = \frac{V(z)}{I(z)} = \frac{V_+ e^{-j\beta z} + V_- e^{j\beta z}}{(V_+ e^{-j\beta z} - V_- e^{j\beta z})Y_0} = Z_0 \frac{1 + \Gamma_L e^{2j\beta z}}{1 - \Gamma_L e^{2j\beta z}} \tag{1.47}$$

$$Z_{in}(-l) = Z_L \frac{\cos\beta l + j\dfrac{Z_0}{Z_L}\sin\beta l}{\cos\beta l + j\dfrac{Z_L}{Z_0}\sin\beta l} \tag{1.48}$$

The voltage and current as function of z may be written as

$$V(z) = V_+ e^{-j\beta z}(1 + \Gamma(z))$$
$$I(z) = Y_0 V_+ e^{-j\beta z}(1 - \Gamma(z)) \tag{1.49}$$

The ratio between the maximum and minimum voltage along a transmission line is called the voltage standing wave ratio S, VSWR.

$$S = \left|\frac{V(z)\,max}{V(z)\,min}\right| = \frac{1 + |\Gamma(z)|}{1 - |\Gamma(z)|} \tag{1.50}$$

1.4 MATCHING TECHNIQUES

Usually in communication systems the load impedance is not the same as the impedance of commercial transmission lines. We will get maximum power transfer from the source to the load if impedances are matched [1–8]. A perfect impedance match corresponds to a VSWR of 1:1. A reflection coefficient magnitude of zero is a perfect match; a value of 1 is perfect reflection. The reflection coefficient (Γ) of a short circuit has a value of −1 (1 at an angle of 180°). The reflection coefficient of an open circuit is 1 at an angle of 0°. The return loss of a load is merely the magnitude of the reflection coefficient expressed in decibels. The correct equation for return loss is Return loss = −20 × log [mag(Γ)].

For a maximum voltage V_m in a transmission line the maximum power will be

$$P_{max} = \frac{V_m^2}{Z_0} \tag{1.51}$$

For an unmatched transmission line the maximum power will be

$$P_{max} = \frac{\left(1-|\Gamma|^2\right)V_+^2}{Z_0}$$

$$V_{max} = (1+|\Gamma|)V_+ \tag{1.52}$$

$$V_+^2 = \frac{V_{max}^2}{(1+|\Gamma|)^2}$$

$$P_{max} = \frac{1-|\Gamma|}{1+|\Gamma|} \cdot \frac{V_{max}^2}{Z_0} = \frac{V_{max}^2}{VSWR \cdot Z_0} \tag{1.53}$$

A 2:1 VSWR will result in half of the maximum power transferred to the load. The reflected power may cause damage to the source.

Equation 1.47 indicates that there is a one-to-one correspondence between the reflection coefficient and input impedance. A movement of distance z along a transmission line corresponds to a change of $e^{-2j\beta z}$, which represents a rotation via an angle of $2\beta z$. In the reflection coefficient plane we may represent any normalized impedance by contours of constant resistance, r, and contours of constant reactance, x. The corresponding impedance moves on a constant radius circle via an angle of $2\beta z$ to a new impedance value. Those relations may be demonstrated by a graphical aid called a Smith chart (Figure 1.5) and represented by the following set of equations:

$$\Gamma_L = \frac{\dfrac{Z_L}{Z_0} - 1}{\dfrac{Z_L}{Z_0} + 1}$$

$$z(l) = \frac{Z_L}{Z_0} = r + jx \tag{1.54}$$

$$\Gamma_L = \frac{r+jx-1}{r+jx+1} = p + jq$$

$$\frac{Z(z)}{Z_0} = \frac{1+\Gamma(z)}{1-\Gamma(z)} = r + jx$$

$$\Gamma(z) = u + jv \tag{1.55}$$

$$\frac{1+u+jv}{1-u-jv} = r + jx$$

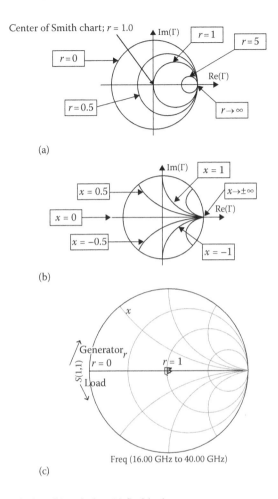

FIGURE 1.5 (a) r circles. (b) x circles. (c) Smith chart.

$$\left(u - \frac{r}{1+r}\right)^2 + v^2 = \frac{1}{(1+r)^2}$$

$$(u-1)^2 + \left(v - \frac{1}{x}\right)^2 = \frac{1}{x^2}$$

(1.56)

Equation 1.56 presents two families of circles in the reflection coefficient plane. The first family comprises contours of constant resistance, r, and the second family comprises contours of constant reactance, x. The center of the Smith chart is $r = 1$. Moving away from the load corresponds to moving around the chart in a clockwise direction. Moving away from the generator toward the load corresponds to moving around the chart in a counterclockwise direction. A complete revolution around the chart in a clockwise direction corresponds to a movement of a half wavelength away from the load. The Smith chart may be employed to calculate the reflection coefficient and the input impedance for a given transmission line and load impedance. If we are at a matched impedance condition at the center of the Smith chart, any length of transmission line with impedance Z_0 does nothing to the input match. But if the reflection coefficient of the network (S_{11}) is at some nonideal impedance, adding transmission line between the network and the reference plane rotates the observed reflection coefficient clockwise about the center of the Smith chart. Further, the rotation occurs at a fixed radius (and VSWR magnitude) if the transmission line has the same characteristic impedance as the source impedance Z_0. Adding a quarter-wavelength means a 180° phase rotation. Adding one quarter-wavelength from a short circuit moves us 180° to the right side of the chart, to an open circuit.

1.4.1 SMITH CHART GUIDELINES

The Smith chart contains almost all possible complex impedances within one circle. The horizontal centerline represents resistance/conductance. Zero resistance is located on the left end of the horizontal centerline. Infinite resistance is located on the right end of the horizontal centerline. Impedances in the Smith chart are normalized to the characteristic impedance of the transmission line and are independent of the characteristic impedance of the transmission. The center of the line and also of the chart is 1.0 point, where $R = Z_0$ or $G = Y_0$.

At point $r = 1.0$, $Z = Z_0$ and no reflection will occur.

1.4.2 QUARTER-WAVE TRANSFORMERS

A quarter-wave transformer may be used to match a device with impedance Z_L to a system with impedance Z_0, as shown in Figure 1.6. A quarter-wave transformer is a matching network with bandwidth somewhat inversely proportional to the relative mismatch we are trying to match. For a single-stage quarter-wave transformer, the correct transformer impedance is the geometric mean between the impedances of the load and the source. If we substitute in Equation 1.47 $l = \dfrac{\lambda}{4}$, $\beta = \dfrac{2\pi}{\lambda}$ we get

$$\bar{Z}(-l) = \frac{\dfrac{Z_L}{Z_{02}}\cos\beta\dfrac{\lambda}{4} + j\sin\beta\dfrac{\lambda}{4}}{\cos\beta\dfrac{\lambda}{4} + j\dfrac{Z_L}{Z_{02}}\sin\beta\dfrac{\lambda}{4}} = \frac{Z_{02}}{Z_L}$$

(1.57)

$$\bar{Z}(-l) = \frac{Z_{01}}{Z_{02}} = \frac{Z_{02}}{Z_L}$$

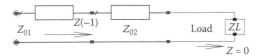

FIGURE 1.6   Quarter-wave transformer.

We will achieve matching when

$$Z_{02} = \sqrt{Z_L Z_{01}} \qquad (1.58)$$

For complex Z_L values Z_{02} also will be a complex impedance. However, standard transmission lines have real impedance values. To match a complex $Z_L = R + jX$ we transform Z_L to a real impedance $Z_{L1} - jX$ to Z_L. Connecting a capacitor $-jX$ to Z_L is not practical at high frequencies. A capacitor at high frequencies has parasitic inductance and resistance. A practical method to transform Z_L to a real impedance Z_{L1} is to add a transmission line with impedance Z_0 and length l to get a real value Z_{L1}.

1.4.3 WIDEBAND MATCHING—MULTISECTION TRANSFORMERS

Multisection quarter-wave transformers are employed for wideband applications. Responses such as Chebyshev (equi-ripple) and maximally flat are possible for multisection transformers. Each section brings us to intermediate impedance. In four section transformers from 25 ohms to 50 ohms intermediate impedances are chosen by using an arithmetic series. For an arithmetic series the steps are equal, $\Delta Z = 6.25\ \Omega$, so the impedances are 31.25 Ω, 37.5 Ω, 43.75 Ω. Solving for the transformers yields $Z_1 = 27.951$, $Z_2 = 34.233$, $Z_3 = 40.505$, and $Z_4 = 46.771\ \Omega$. A second solution to multisection transformers involves a geometric series from impedance Z_L to impedance Z_S. Here the impedance from one section to the next adjacent section is a constant ratio.

1.4.4 SINGLE-STUB MATCHING

A device with admittance Y_L can be matched to a system with admittance Y_0 by using a shunt or series single stub (Figure 1.7). At a distance l from the load we can get a normalized admittance $\overline{Y}_{in} = 1 + j\overline{B}$. By solving Equation 1.58 we can calculate l.

$$\overline{Y}(l) = \frac{1 + j \dfrac{Z_L}{Z_0} \tan \beta l}{\dfrac{Z_L}{Z_0} + j \tan \beta l} = 1 + j\overline{B} \qquad (1.59)$$

At this location we can add a shunt stub with normalized input susceptance, $-j\overline{B}$, to yield $\overline{Y}_{in} = 1$ as presented in Equation 1.60. $\overline{Y}_{in} = 1$ represents a matched load. The

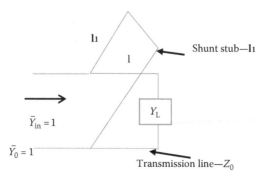

FIGURE 1.7 Single-stub matching.

stub can be a short circuited line or open circuited line. The susceptance $\bar{B}$ is given in Equation 1.60.

$$\bar{Y}_{in} = 1 + j\bar{B}$$
$$\bar{Y}_{1n} = -j\bar{B} \qquad (1.60)$$
$$\bar{B} = ctg\beta l_1$$

The length l_1 of the short circuited line may be calculated by solving Equation 1.61.

$$Im\left\{ \frac{1 + j\dfrac{Z_L}{Z_0}\tan\beta l}{\dfrac{Z_L}{Z_0} + j\tan\beta l} \right\} = ctg\beta l_1 \qquad (1.61)$$

1.5 COAXIAL TRANSMISSION LINE

A coaxial transmission line consists of two round conductors in which one completely surrounds the other, with the two separated by a continuous solid dielectric [1–6]. The desired propagation mode is TEM. The major advantage of a coaxial over a microstrip line is that the transmission line does not radiate. The disadvantages are that coaxial lines are more expensive. Coaxial line are usually employed up to 18 GHz. Coaxial lines are very expensive at frequencies higher than 18 GHz. To obtain good performance at higher frequencies, small-diameter cables are required to stay below the cutoff frequency. Maxwell laws are employed to compute the electric and magnetic fields in the coaxial transmission line. A cross section of a coaxial transmission line is shown in Figure 1.8.

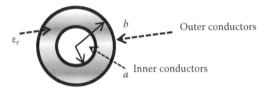

FIGURE 1.8 Coaxial transmission line.

$$\oint_S E \cdot ds = \frac{1}{\varepsilon_0} \int_V \rho \, dv = \frac{1}{\varepsilon_0} q \tag{1.62}$$

$$E_r = \frac{\rho_L}{2\pi r \varepsilon}$$

$$V = \int_a^b E \cdot dr = \int_a^b \frac{\rho_L}{2\pi r \varepsilon} \cdot dr = \frac{\rho_L}{2\pi \varepsilon} \ln \frac{b}{a} \tag{1.63}$$

Ampère's law is employed to calculate the magnetic field.

$$\oint H \cdot dl = 2\pi r H_\phi = I$$

$$H_\phi = \frac{I}{2\pi r} \tag{1.64}$$

$$V = \int_a^b E_r \cdot dr = -\int_a^b \eta H_\phi \cdot dr = \frac{I\eta}{2\pi} \ln \frac{b}{a} \tag{1.65}$$

$$Z_0 = \frac{V}{I} = \frac{\eta}{2\pi} \ln \frac{b}{a} \qquad \eta = \sqrt{\mu/\varepsilon}$$

The power flow in the coaxial transmission line may be calculated by calculating the Poynting vector.

$$P = (E \times H) \cdot n = EH$$

$$P = \frac{VI}{2\pi r^2 \ln(b/a)} \tag{1.66}$$

$$W = \int_s P \cdot ds = \int_a^b \frac{VI}{2\pi r^2 \ln(b/a)} 2\pi r \, dr = VI$$

TABLE 1.3

Industry-Standard Coax Cables

Cable Type	Outer Diameter (inches)	2b (inches)	2a (inches)	Z_0 (ohms)	f_c (GHz)
RG-8A	0.405	0.285	0.089	50	14.0
RG-58A	0.195	0.116	0.031	50	35.3
RG-174	0.100	0.060	0.019	50	65.6
RG-196	0.080	0.034	0.012	50	112
RG-214	0.360	0.285	0.087	50	13.9
RG-223	0.216	0.113	0.037	50	34.6
SR-085	0.085	0.066	0.0201	50	60.2
SR-141	0.141	0.1175	0.0359	50	33.8
SR-250	0.250	0.210	0.0641	50	18.9

Table 1.3 presents several industry-standard coaxial cables. The cables' dimensions, impedance, and cutoff frequency are given in Table 1.3. RG cables are flexible cables. SR cables are semirigid cables.

1.5.1 Cutoff Frequency and Wavelength of Coaxial Cables

The criterion for cutoff frequency (f_c) is that the circumference at the midpoint inside the dielectric must be less than a wavelength. Therefore the cutoff wavelength for the TE_{01} mode is $\lambda_c = \pi(a+b)\sqrt{\mu_r \varepsilon_r}$.

1.6 MICROSTRIP LINE

A microstrip is a planar printed transmission line. A microstrip has been the most popular radio frequency transmission line over the last 20 years [3,7]. Microstrip transmission lines consist of a conductive strip of width W and thickness t and a wider ground plane, separated by a dielectric layer of thickness H. In practice, a microstrip line is usually made by etching circuitry on a substrate that has a ground plane on the opposite face. A cross section of the microstrip line is shown in Figure 1.9. The major advantage of a microstrip over a stripline is that all components can be mounted on top of the board. The disadvantages are that when high isolation is required such as in a filter or switch, some external shielding is needed. Microstrip circuits may

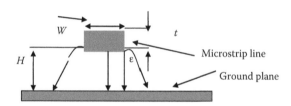

FIGURE 1.9   Microstrip line cross section.

radiate, causing an unintended circuit response. A microstrip is dispersive; signals of different frequencies travel at slightly different speeds. Other microstrip line configurations are offset stripline and suspended air microstrip line. For a microstrip line not all of the fields are constrained to the same dielectric. At the line edges the fields pass via air and a dielectric substrate. The effective dielectric constant should be calculated.

1.6.1 Effective Dielectric Constant

Part of the fields in the microstrip line structure exists in air and the other part of the fields exists in the dielectric substrate. The effective dielectric constant is somewhat less than the substrate's dielectric constant. The effective dielectric constant of the microstrip line is calculated by

$$\text{For } \left(\frac{W}{H}\right) < 1 \tag{1.67}$$

$$\varepsilon_e = \frac{\varepsilon_r + 1}{2} + \frac{\varepsilon_r - 1}{2}\left[\left(1 + 12\left(\frac{H}{W}\right)\right)^{-0.5} + 0.04\left(1 - \left(\frac{W}{H}\right)\right)^2\right]$$

$$\text{For } \left(\frac{W}{H}\right) \geq 1 \tag{1.68}$$

$$\varepsilon_e = \frac{\varepsilon_r + 1}{2} + \frac{\varepsilon_r - 1}{2}\left[\left(1 + 12\left(\frac{H}{W}\right)\right)^{-0.5}\right]$$

This calculation ignores strip thickness and frequency dispersion, but their effects are negligible.

1.6.2 Characteristic Impedance

The characteristic impedance Z_0 is a function of the ratio of the height to the width W/H of the transmission line, and also has separate solutions depending on the value of W/H. The characteristic impedance Z_0 of a microstrip is calculated by

$$\text{For } \left(\frac{W}{H}\right) < 1 \tag{1.69}$$

$$Z_0 = \frac{60}{\sqrt{\varepsilon_e}}\ln\left[8\left(\frac{H}{W}\right) + 0.25\left(\frac{H}{W}\right)\right]\Omega$$

$$\text{For } \left(\frac{W}{H}\right) \geq 1 \tag{1.70}$$

$$Z_0 = \frac{120\pi}{\sqrt{\varepsilon_e}\left[\left(\frac{H}{W}\right) + 1.393 + 0.66 * \ln\left(\frac{H}{W} + 1.444\right)\right]} \, \Omega$$

We can calculate Z_0 by using Equations 1.69 and 1.70 for a given $\left(\frac{W}{H}\right)$. However, to calculate $\left(\frac{W}{H}\right)$ for a given Z_0 we first should calculate ε_e. However, to calculate ε_e we should know $\left(\frac{W}{H}\right)$. We first assume that $\varepsilon_e = \varepsilon_r$ and compute $\left(\frac{W}{H}\right)$ for this value of $\left(\frac{W}{H}\right)$ we compute ε_e. Then we compute a new value of $\left(\frac{W}{H}\right)$. Two to three iterations are needed to calculate accurate values of $\left(\frac{W}{H}\right)$ and ε_e. We may calculate $\left(\frac{W}{H}\right)$ with around 10% accuracy by using Equation 1.71.

$$\frac{W}{H} = 8 \frac{\sqrt{\left[e^{\frac{Z_0}{42.4}\sqrt{(\varepsilon_r+1)}} - 1\right]}\left[\dfrac{7 + \dfrac{4}{\varepsilon_r}}{11}\right] + \left[\dfrac{1 + \dfrac{1}{\varepsilon_r}}{0.81}\right]}{\left[e^{\frac{Z_0}{42.4}\sqrt{(\varepsilon_r+1)}} - 1\right]} \tag{1.71}$$

1.6.3 Higher-Order Transmission Modes in a Microstrip Line

To prevent higher-order transmission modes we should limit the thickness of the microstrip substrate to 10% of a wavelength. The cutoff frequency of the higher-order transmission mode is given as $f_c = \dfrac{c}{4H\sqrt{\varepsilon-1}}$.

1.6.3.1 Examples

Higher order modes will not propagate in a microstrip lines printed on Alumina substrate 15 mil thick up to 18 GHz. Higher order modes will not propagate in a microstrip lines printed on GaAs substrate 4 mil thick up to 80 GHz. Higher order

TABLE 1.4

Examples of Microstrip Line Parameters

Substrate	W/H	Impedance Ω
Alumina ($\varepsilon_r = 9.8$)	0.95	50
GaAs ($\varepsilon_r = 12.9$)	0.75	50
$\varepsilon_r = 2.2$	3	50

modes will not propagate in a microstrip lines printed on quartz substrate 5 mil thick up to 120 GHz. Table 1.4 presents several examples of microstrip line parameters.

1.6.3.2 Losses in Microstrip Line

Losses in microstrip line are due to conductor loss, radiation loss, and dielectric loss.

1.6.4 CONDUCTOR LOSS

Conductor loss may be calculated by using Equation 1.72.

$$\alpha_c = 8.686 \log(R_S/(2WZ_0)) \quad \text{dB/Length}$$
$$R_S = \sqrt{\pi f \mu \rho} \quad \text{Skin resistance}$$

(1.72)

Conductor losses may also be calculated by defining an equivalent loss tangent δ_c, given by $\delta_c = \delta_s/h$, where $\delta_s = \sqrt{\dfrac{2}{\omega \mu \sigma}}$. Where σ is the strip conductivity, h is the substrate height and μ is the free space permeability.

1.6.5 DIELECTRIC LOSS

Dielectric loss may be calculated by using Equation 1.73.

$$\alpha_d = 27.3 \frac{\varepsilon_r}{\sqrt{\varepsilon_{\text{eff}}}} \frac{\varepsilon_{\text{eff}} - 1}{\varepsilon_r - 1} \frac{tg\delta}{\lambda_0} \quad \text{dB/cm}$$

(1.73)

$$tg\delta = \text{dielectric loss coefficient}$$

Losses in microstrip lines are presented in Tables 1.5 to 1.7 for several microstrip line structures. For example, total loss of a microstrip line presented in Table 1.6 at 40 GHz is 0.5 dB/cm. Total loss of a microstrip line presented in Table 1.7 at 40 GHz is 1.42 dB/cm. We may conclude that losses in microstrip lines limit the applications of microstrip technology at millimeter wave frequencies.

TABLE 1.5
Microstrip Line Losses for Alumina Substrate 10 Mil Thick

Frequency (GHz)	Tangent Loss (dB/cm)	Metal Loss (dB/cm)	Total Loss (dB/cm)
10	−0.005	−0.12	−0.124
20	−0.009	−0.175	−0.184
30	−0.014	−0.22	−0.23
40	−0.02	−0.25	−0.27

Note: Line parameters: Alumina, $H = 254$ μm (10 mils), $W = 247$ μm, $E_r = 9.9$, tan δ = 0.0002, 3 μm gold, conductivity 3.5 E7 mhos/m.

TABLE 1.6
Microstrip Line Losses for Alumina Substrate 5 Mil Thick

Frequency (GHz)	Tangent Loss (dB/cm)	Metal Loss (dB/cm)	Total Loss (dB/cm)
10	−0.004	−0.23	−0.23
20	−0.009	−0.333	−0.34
30	−0.013	−0.415	−0.43
40	−0.018	−0.483	−0.5

Note: Alumina, $H = 127$ μm (5 mils), $W = 120$ μm, $E_r = 9.9$, tan δ = 0.0002, 3 μm gold, conductivity 3.5 E7 mhos/m.

TABLE 1.7
Microstrip Line Losses for GaAs Substrate 2 Mil Thick

Frequency (GHz)	Tangent Loss (dB/cm)	Metal Loss (dB/cm)	Total Loss (dB/cm)
10	−0.010	−0.66	−0.67
20	−0.02	−0.96	−0.98
30	−0.03	−1.19	−1.22
40	−0.04	−1.38	−1.42

Note: GaAs, $H = 50$ μm (2 mils), $W = 34$ μm, $E_r = 12.88$, tan δ = 0.0004, 3 μm gold, conductivity 3.5 E7 mhos/m.

TABLE 1.8
Hard Materials—Ceramics

Material	Symbol or Formula	Dielectric Constant	Dissipation Factor (tan δ)	Coefficient of Thermal Expansion (ppm/Â°C)	Thermal Cond (W/mÂ°C)	Mass Density (gr/cc)
Alumina 99.5%	Al_2O_3	9.8	0.0001	8.2	35	3.97
Alumina 96%	Al_2O_3	9.0	0.0002	8.2	24	3.8
Aluminum nitride	AlN	8.9	0.0005	7.6	290	3.26
Beryllium oxide	BeO	6.7	0.003	6.05	250	
Gallium arsenide	GaAs	12.88	0.0004	6.86	46	5.32
Indium phosphide	InP	12.4				
Quartz		3.8	0.0001	0.6	5	
Sapphire		9.3, 11.5				
Silicon (high resistivity)	Si (HRS)	4		2.5	138	2.33
Silicon carbide	SiC	10.8	0.002	4.8	350	3.2

TABLE 1.9
Soft Materials

Manufacturer and Material	Symbol or Formula	Relative Dielectric Constant ε_r	Tolerance on Dielectric Constant	Tan δ	Mass Density (gr/cc)	Thermal Conductivity (W/m°C)	Coefficient of Thermal Expansion PPM/°C x/y/z
Rogers Duroid 5870	PTFE/random glass	2.33	0.02	0.0012	2.2	0.26	22/28/173
Rogers Duroid 5880	PTFE/random glass	2.2	0.02	0.0012	2.2	0.26	31/48/237
Rogers Duroid 6002	PTFE/random glass	2.94	0.04	0.0012	2.1	0.44	16/16/24
Rogers Duroid 6006	PTFE/random glass	6	0.15	0.0027	2.7	0.48	38/42/24
Rogers Duroid 6010	PTFE/random glass	10.2–10.8	0.25	0.0023	2.9	0.41	24/24/24
FR-4	Glass/epoxy	4.8		0.022		0.16	
Polyethylene		2.25					
Polyflon CuFlon	PTFE	2.1		0.00045			12.9
Polyflon PolyGuide	Polyolefin	2.32		0.0005			108
Polyflon Norclad	Thermoplastic	2.55		0.0011			53
Polyflon Clad Ultem	Thermoplastic	3.05		0.003	1.27		56
PTFE	PTFE	2.1		0.0002	2.1	0.2	
Rogers R/flex 3700	Thermally stable thermoplastic	2.0		0.002			8
Rogers RO3003	PTFE ceramic	3	0.04	0.0013	2.1	0.5	17/17/24
Rogers RO3006	PTFE ceramic	6.15	0.15	0.0025	2.6	0.61	17/17/24
Rogers RO3010	PTFE ceramic	10.2	0.3	0.0035	3	0.66	17/17/24
Rogers RO3203	PTFE ceramic	3.02					
Rogers RO3210	PTFE ceramic	10.2					
Rogers RO4003	Thermoset plastic ceramic glass	3.38	0.05	0.0027	1.79	0.64	11/14/46
Rogers RO4350B	Thermoset plastic ceramic glass	3.48	0.05	0.004	1.86	0.62	14/16/50
Rogers TMM 3	Ceramic/thermoset	3.27	0.032	0.002	1.78	0.7	15/15/23
Rogers TMM 10	Ceramic/thermoset	9.2	0.23	0.0022	2.77	0.76	21/21/20

1.7 MATERIALS

In Table 1.8 hard materials are presented. Alumina is the most popular hard substrate in microwave integrated circuits (MIC). GaAs is the most popular hard substrate in monolithic microwave integrated circuits (MIMIC) technology at microwave frequencies.

In Table 1.9 soft materials are presented. Duroid® is the most popular soft substrate in MIC circuits and in the printed antenna industry. Dielectric losses in Duroid are significantly lower than dielectric losses in an FR-4 substrate. However, the cost of FR-4 substrate is significantly lower than the cost of Duroid. Commercial MIC devices use usually FR-4 substrate. Duroid is the most popular soft substrate used in the development of printed antennas with high efficiency at microwave frequencies.

1.8 WAVEGUIDES

Waveguides are low-loss transmission lines. Waveguides may be rectangular or circular. A rectangular waveguide structure is presented in Figure 1.10. Waveguide structure is uniform in the z direction. Fields in waveguides are evaluated by solving the Helmholtz equation. A wave equation is given in Equation 1.74. A wave equation in a rectangular coordinate system is given in Equation 1.75.

$$\nabla^2 E = \omega^2 \mu \varepsilon E = -k^2 E$$
$$\nabla^2 H = \omega^2 \mu \varepsilon H = -k^2 H \tag{1.74}$$
$$k = \omega \sqrt{\mu \varepsilon} = \frac{\omega}{v} = \frac{2\pi}{\lambda}$$

$$\nabla^2 E = \frac{\partial^2 E_i}{\partial x^2} + \frac{\partial^2 E_i}{\partial y^2} + \frac{\partial^2 E_i}{\partial z^2} + k^2 E_i = 0 \quad i = x, y, z$$

$$\nabla^2 H = \frac{\partial^2 H_i}{\partial x^2} + \frac{\partial^2 H_i}{\partial y^2} + \frac{\partial^2 H_i}{\partial z^2} + k^2 H_i = 0 \tag{1.75}$$

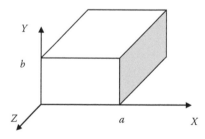

FIGURE 1.10   Rectangular waveguide structure.

The wave equation solution may be written as $E = f(z)g(x,y)$. Field variation in the z direction may be written as $e^{-j\beta z}$. The derivative of this expression in the z direction may be written as $-j\beta e^{-j\beta z}$. Maxwell equations may be presented as written in Equations 1.76 and 1.77. A field may be represented as a superposition of waves in the transverse and longitudinal directions.

$$E(x,y,z) = e(x,y)e^{-j\beta z} + e_z(x,y)e^{-j\beta z}$$

$$H(x,y,z) = h(x,y)e^{-j\beta z} + h_z(x,y)e^{-j\beta z}$$

$$\nabla \times E = (\nabla_t - j\beta a_z) \times (e + e_z)e^{-j\beta z} = -j\omega\mu(h + h_z)e^{-j\beta z}$$

$$\nabla_t \times e - j\beta a_z \times e + \nabla_t \times e_z - j\beta a_z \times e_z = -j\omega\mu(h + h_z) \qquad (1.76)$$

$$a_z \times e_z = 0$$

$$\nabla_t \times e = -j\omega\mu h_z$$

$$-j\beta a_z \times e + \nabla_t \times e_z = -a_z \times \nabla_t e_z - j\beta a_z \times e = -j\omega\mu h$$

$$\nabla_t \times h = -j\omega\varepsilon e_z$$

$$a_z \times \nabla_t h_z + j\beta a_z \times h = -j\omega\varepsilon e \qquad (1.77)$$

$$\nabla \cdot \mu H = (\nabla_t - j\beta a_z) \cdot (h + h_z)\mu e^{-j\beta z} = 0$$

$$\nabla_t \cdot h - j\beta a_z \cdot h_z = 0 \qquad (1.78)$$

$$\nabla \cdot \varepsilon E = 0 \quad \nabla_t \cdot e - j\beta a_z \cdot e_z = 0$$

Waves may be characterized as TEM, TE, or TM waves. In TEM waves $e_z = h_z = 0$. In TE waves $e_z = 0$. In TM waves $h_z = 0$.

1.8.1   TE WAVES

In TE waves $e_z = 0$. h_z is given as the solution of Equation 1.79. The solution to Equation 1.79 may be written as $h_z = f(x)g(y)$.

$$\nabla_t^2 h_z = \frac{\partial^2 h_z}{\partial x^2} + \frac{\partial^2 h_z}{\partial y^2} + k_c^2 h_z = 0 \qquad (1.79)$$

By applying $h_z = f(x)g(y)$ to Equation 1.79 and dividing by fg we get Equation 1.80.

$$\frac{f''}{f} + \frac{g''}{g} + k_c^2 = 0 \qquad (1.80)$$

f is a function that varies in the x direction and g is a function that varies in the y direction. The sum of f and g may be equal to zero only if they equal a constant.

These facts are written in Equation 1.81.

$$\frac{f''}{f} = -k_x^2; \frac{g''}{g} = -k_y^2$$

$$k_x^2 + k_y^2 = k_c^2$$

(1.81)

The solutions for f and g are given in Equation 1.82. A_1, A_2, B_1, B_2 are derived by applying h_z boundary conditions to Equation 1.80.

$$f = A_1 \cos k_x x + A_2 \sin k_x x$$
$$g = B_1 \cos k_y y + B_2 \sin k_y y$$

(1.82)

h_z boundary conditions are written in Equation 1.83.

$$\frac{\partial h_z}{\partial x} = 0 \quad @ \quad x = 0, a$$

$$\frac{\partial h_z}{\partial y} = 0 \quad @ \quad y = 0, b$$

(1.83)

By applying h_z boundary conditions to Equation 1.82 we get the relations written in Equation 1.84.

$$-k_x A_1 \sin k_x x + k_x A_2 \cos k_x x = 0$$

$$-k_y B_1 \sin k_y y + k_y B_2 \cos k_y y = 0$$

$$A_2 = 0 \quad k_x a = 0 \quad k_x = \frac{n\pi}{a} \quad n = 0, 1, 2$$

(1.84)

$$B_2 = 0 \quad k_y b = 0 \quad k_x = \frac{m\pi}{b} \quad m = 0, 1, 2$$

The solution for h_z is given in Equation 1.85.

$$h_z = A_{nm} \cos \frac{n\pi x}{a} \cos \frac{m\pi y}{b}$$

$$n = m \neq 0 \quad n = 0, 1, 2 \quad m = 0, 1, 2$$

(1.85)

$$k_{c,nm} = \left[\left(\frac{n\pi}{a} \right)^2 + \left(\frac{m\pi}{b} \right)^2 \right]^{1/2}$$

Both n and m cannot be zero. The wave number at cutoff is $k_{c,nm}$ and depends on the waveguide dimensions. The propagation constant γ_{nm} is given in Equation 1.86.

$$\gamma_{nm} = j\beta_{nm} = j(k_0^2 - k_c^2)^{1/2}$$

$$= j\left[\left(\frac{2\pi}{\lambda_0}\right)^2 - \left(\frac{n\pi}{a}\right)^2 - \left(\frac{m\pi}{b}\right)^2\right]^{1/2} \qquad (1.86)$$

For $k_0 > k_{c,nm}$, β is real and the wave will propagate. For $k_0 < k_{c,nm}$, β is imaginary and the wave will decay rapidly. Frequencies that define propagating and decaying waves are called cutoff frequencies. We may calculate cutoff frequencies by using Equation 1.87.

$$f_{c,nm} = \frac{c}{2\pi} k_{c,nm} = \frac{c}{2\pi}\left[\left(\frac{n\pi}{a}\right)^2 + \left(\frac{m\pi}{b}\right)^2\right]^{1/2} \qquad (1.87)$$

For $a = 2$ cutoff wavelength is computed by using Equation 1.88.

$$\lambda_{c,nm} = \frac{2ab}{[n^2 b^2 + m^2 a^2]^{1/2}} = \frac{2a}{[n^2 + 4m^2]^{1/2}}$$

$$\lambda_{c,10} = 2a \quad \lambda_{c,01} = a \quad \lambda_{c,11} = 2a/\sqrt{5} \qquad (1.88)$$

$$\frac{c}{2a} \langle f_{01}\rangle \frac{c}{a}$$

For $\dfrac{c}{2a} \langle f_{01}\rangle \dfrac{c}{a}$ the dominant is TE_{10}.

By using Equations 1.76 to 1.78 we can derive the electromagnetic fields that propagate in the waveguide as given in Equation 1.89.

$$H_z = A_{nm} \cos\frac{n\pi x}{a} \cos\frac{m\pi y}{b} e^{\pm j\beta_{nm} z}$$

$$H_x = \pm j\frac{n\pi\beta_{nm}}{ak_{c,nm}^2} A_{nm} \sin\frac{n\pi x}{a} \cos\frac{m\pi y}{b} e^{\pm j\beta_{nm} z}$$

$$H_y = \pm j\frac{m\pi\beta_{nm}}{bk_{c,nm}^2} A_{nm} \cos\frac{n\pi x}{a} \sin\frac{m\pi y}{b} e^{\pm j\beta_{nm} z} \qquad (1.89)$$

$$E_X = Z_{h,nm} j\frac{m\pi\beta_{nm}}{bk_{c,nm}^2} A_{nm} \cos\frac{n\pi x}{a} \sin\frac{m\pi y}{b} e^{\pm j\beta_{nm} z}$$

$$E_Y = -jZ_{h,nm}\frac{n\pi\beta_{nm}}{ak_{c,nm}^2} A_{nm} \sin\frac{n\pi x}{a} \cos\frac{m\pi y}{b} e^{\pm j\beta_{nm} z}$$

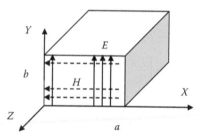

FIGURE 1.11 TE_{10} mode.

The impedance of the *nm* modes is given as $Z_{h,nm} = \dfrac{e_x}{h_y} = \dfrac{k_0}{\beta_{nm}} \sqrt{\dfrac{\mu_0}{\varepsilon_0}}$. The power of the *nm* mode is computed by using Poynting vector calculation as shown in Equation 1.90.

$$P_{nm} = 0.5\,\mathrm{Re} \int_0^a \int_0^b E \times H^* \cdot a_z \, dx\, dy = 0.5\,\mathrm{Re}\, Z_{h,nm} \int_0^a \int_0^b \left(H_y H_Y^* + H_x H_x^* \right) dx\, dy$$

$$\int_0^a \int_0^b \cos^2 \frac{n\pi x}{a} \sin^2 \frac{m\pi y}{b} \, dx\, dy = \frac{ab}{4} \quad n \neq 0 \quad m \neq 0$$

(1.90)

or $\quad \dfrac{ab}{2} \quad$ nor $\quad m = 0$

$$P_{nm} = \frac{|A_{nm}|^2}{2\varepsilon_{0n}\varepsilon_{0m}} \left(\frac{\beta_{nm}}{k_{c,nm}} \right)^2 Z_{h,nm} \quad \varepsilon_{0n} = 1,\ n = 0 \quad \varepsilon_{0n} = 2,\ n \succ 0$$

The TE mode with the lowest cutoff frequency in rectangular waveguide is TE_{10}. TE_{10} fields in a rectangular waveguide are shown in Figure 1.11.

1.8.2   TM Waves

In TM waves, $h_z = 0$. e_z is given as the solution of Equation 1.91. The solution to Equation 1.91 may be written as $e_z = f(x)g(y)$. e_z should be zero at the metallic walls. e_z boundary conditions are written in Equation 1.92. The solution for e_z is given in Equation 1.93.

$$\nabla_t^2 e_z = \frac{\partial^2 e_z}{\partial x^2} + \frac{\partial^2 e_z}{\partial y^2} + k_c^2 e_z = 0$$

(1.91)

$$e_z = 0 \quad @ \quad x = 0, a$$
$$e_z = 0 \quad @ \quad y = 0, b$$

(1.92)

$$e_z = A_{nm} \sin \frac{n\pi x}{a} \sin \frac{m\pi y}{b}$$

$$n = m \neq 0 \quad n = 0,1,2 \quad m = 0,1,2 \tag{1.93}$$

$$k_{c,nm} = \left[\left(\frac{n\pi}{a} \right)^2 + \left(\frac{m\pi}{b} \right)^2 \right]^{1/2}$$

The first propagating TM mode is TM_{11}, $n = m = 1$. By using Equations 1.76 to 1.78 and 1.91 we can derive the electromagnetic fields that propagate in the waveguide as given in Equation 1.94.

$$E_z = \sin \frac{n\pi x}{a} \sin \frac{m\pi y}{b} e^{\pm j\beta_{nm} z}$$

$$E_x = -j \frac{n\pi \beta_{nm}}{a k_{c,nm}^2} \cos \frac{n\pi x}{a} \sin \frac{m\pi y}{b} e^{\pm j\beta_{nm} z}$$

$$E_y = -j \frac{m\pi \beta_{nm}}{b k_{c,nm}^2} A_{nm} \sin \frac{n\pi x}{a} \cos \frac{m\pi y}{b} e^{\pm j\beta_{nm} z} \tag{1.94}$$

$$H_X = \frac{-E_y}{Z_{e,nm}}$$

$$H_Y = \frac{E_x}{Z_{e,nm}}$$

The impedance of the *nm* modes is $Z_{e,nm} = \frac{\beta_{nm}}{k_0} \sqrt{\frac{\mu_0}{\varepsilon_0}}$. The TM mode with the lowest cutoff frequency in rectangular waveguide is TM_{11}. TM_{11} fields in rectangular waveguide are shown in Figure 1.12.

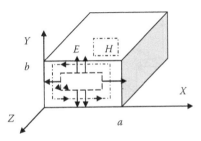

FIGURE 1.12 TM_{11} mode.

1.9 CIRCULAR WAVEGUIDE

A circular waveguide is used to transmit electromagnetic waves in circular polarization. At high frequencies attenuation of several modes in a circular waveguide is lower than in a rectangular waveguide. The circular waveguide structure is uniform in the z direction. Fields in waveguides are evaluated by solving the Helmoltz equation in a cylindrical coordinate system. A circular waveguide in a cylindrical coordinate system is presented in Figure 1.13. A wave equation is given in Equation 1.95. A wave equation in a cylindrical coordinate is given in Equation 1.96.

$$\nabla^2 E = \omega^2 \mu \varepsilon E = -k^2 E$$

$$\nabla^2 H = \omega^2 \mu \varepsilon H = -k^2 H \tag{1.95}$$

$$k = \omega \sqrt{\mu \varepsilon} = \frac{\omega}{v} = \frac{2\pi}{\lambda}$$

$$\nabla^2 E = \frac{1}{r} \frac{\partial}{\partial r} \left(r \frac{\partial E_i}{\partial r} \right) + \frac{1}{r^2} \frac{\partial^2 E_i}{\partial \phi^2} + \frac{\partial^2 E_i}{\partial z^2} - \gamma^2 E_i = 0 \quad i = r, \phi, z$$

$$\nabla^2 H = \frac{1}{r} \frac{\partial}{\partial r} \left(r \frac{\partial H_i}{\partial r} \right) + \frac{1}{r^2} \frac{\partial^2 H_i}{\partial \phi^2} + \frac{\partial^2 E_i}{\partial z^2} - \gamma^2 H_i = 0 \tag{1.96}$$

The solution to Equation 1.96 may be written as $E = f(r)g(\phi)h(z)$. By applying E to Equation 1.96 and dividing by $f(r)g(\phi)h(z)$ we get Equation 1.97.

$$\nabla^2 E = \frac{1}{rf} \frac{\partial}{\partial r} \left(r \frac{\partial f}{\partial r} \right) + \frac{1}{r^2 g} \frac{\partial^2 g}{\partial \phi^2} + \frac{\partial^2 h}{h \partial z^2} - \gamma^2 = 0 \tag{1.97}$$

f is a function that varies in the r direction, g is a function that varies in the ϕ direction, and h is a function that varies in the z direction. The sum of f, g, and h may be equal to zero only if they equal to a constant. The solution for h is written in Equation 1.98. The propagation constant is γg.

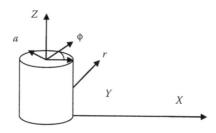

FIGURE 1.13   Circular waveguide structure.

$$\frac{\partial^2 h}{h \partial z^2} - \gamma_g^2 = 0$$

$$h = Ae^{-\gamma_g z} + Be^{\gamma_g z} \tag{1.98}$$

$$\frac{r}{f}\frac{\partial}{\partial r}\left(r\frac{\partial f}{\partial r}\right) + \frac{1}{g}\frac{\partial^2 g}{\partial \phi^2} - \left(\gamma^2 - \gamma_g^2\right) = 0 \tag{1.99}$$

The solution for g is written in Equation 1.100.

$$\frac{\partial^2 g}{g \partial \phi^2} + n^2 = 0 \tag{1.100}$$

$$g = A_n \sin n\phi + B_n \cos n\phi$$

$$r\frac{\partial}{\partial r}\left(r\frac{\partial f}{\partial r}\right) + \left((k_c r)^2 - n^2\right)f = 0 \tag{1.101}$$

$$k_c^2 + \gamma^2 = \gamma_g^2$$

Equation 1.101 is a Bessel equation. The solution of this equation is written in Equation 1.102. $J_n(k_c r)$ is a Bessel equation with order n and represents a standing wave. The wave varies a cosine function in the circular waveguide. $N_n(k_c r)$ is a Bessel equation with order n and represents a standing wave. The wave varies a sine function in the circular waveguide.

$$f = C_n J_n(k_c r) + D_n N_n(k_c r) \tag{1.102}$$

The general solution for the electric fields in the circular waveguide is given in Equation 1.103. For $r = 0$ $N_n(k_c r)$ goes to infinity, so $D_n = 0$.

$$E(r,\phi,z) = (C_n J_n(k_c r) + D_n N_n(k_c r))(A_n \sin n\phi + B_n \cos n\phi)e^{\pm \gamma_g z} \tag{1.103}$$

$$A_n \sin n\phi + B_n \cos n\phi = \sqrt{A_n^2 + B_n^2}\cos\left(n\phi + \tan^{-1}\left(\frac{A_n}{B_n}\right)\right) = F_n \cos n\phi \tag{1.104}$$

The general solution for the electric fields in the circular waveguide is given in Equation 1.105.

$$E(r,\phi,z) = E_0 (J_n(k_c r))(\cos n\phi)e^{\pm \gamma_g z} \quad \text{If} \quad \alpha = 0$$

$$E(r,\phi,z) = E_0 (J_n(k_c r))(\cos n\phi)e^{\pm \beta_g z} \tag{1.105}$$

$$\beta_g = \pm\sqrt{\omega^2 \mu\varepsilon - k_c^2}$$

1.9.1 TE WAVES IN A CIRCULAR WAVEGUIDE

In TE waves $e_z = 0$. H_z is given as the solution of Equation 1.106. The solution to Equation 1.106 may be written as given in Equation 1.107.

$$\nabla^2 H_z = \gamma^2 H_z \tag{1.106}$$

$$H_z = H_{0z}(J_n(k_c r))(\cos n\phi)e^{\pm j\beta_g z} \tag{1.107}$$

The electric and magnetic fields are the solution of Maxwell equations as written in Equations 1.108 and 1.109.

$$\nabla X E = -j\omega\mu H \tag{1.108}$$

$$\nabla \times H = j\omega\varepsilon E \tag{1.109}$$

Field variation in the z direction may be written as $e^{-j\beta z}$. The derivative of this expression in the direction may be written as $-j\beta e^{j\beta z}$. The electric and magnetic field components are solutions of Equations 1.110 and 1.111.

$$
\begin{aligned}
E_r &= -\frac{j\omega\mu}{k_c^2}\frac{1}{r}\left(\frac{\partial H_z}{\partial \phi}\right) \\
E_\phi &= \frac{j\omega\mu}{k_c^2}\left(\frac{\partial H_z}{\partial r}\right)
\end{aligned}
\tag{1.110}
$$

$$
\begin{aligned}
H_\phi &= -\frac{-j\beta_g}{k_c^2}\frac{1}{r}\left(\frac{\partial H_z}{\partial \phi}\right) \\
H_r &= \frac{-j\beta_g}{k_c^2}\left(\frac{\partial H_z}{\partial r}\right)
\end{aligned}
\tag{1.111}
$$

H_z, H_r, and E_ϕ boundary conditions are written in Equation 1.112.

$$
\begin{aligned}
\frac{\partial H_z}{\partial r} &= 0 \quad @ \quad r = a \\
H_r &= 0 \quad @ \quad r = a \\
E_\phi &= 0 \quad @ \quad r = a
\end{aligned}
\tag{1.112}
$$

By applying the boundary conditions to Equation 1.107 we get the relations written in Equation 1.113. The solutions of Equation 1.113 are listed in Table 1.10.

TABLE 1.10

Circular Waveguide TE Modes

P	$(n = 0)$ X'_{np}	$(n = 1)$ X'_{np}	$(n = 2)$ X'_{np}	$(n = 3)$ X'_{np}	$(n = 4)$ X'_{np}	$(n = 5)$ X'_{np}
1	3.832	1.841	3.054	4.201	5.317	6.416
2	7.016	5.331	6.706	8.015	9.282	10.52
3	10.173	8.536	9.969	11.346	12.682	13.987
4	13.324	11.706	13.170			

$$\frac{\partial H_z}{\partial r}\bigg|r = a = H_{0z}\left(J'_n(k_c a)\right)(\cos n\phi)e^{-\beta_g z} = 0$$

$$J'_n(k_c a) = 0 \tag{1.113}$$

The wave number at cutoff is $k_{c,np}$. $k_{c,np}$ depends on the waveguide dimensions. The propagation constant $\gamma_{g,np}$ is given in Equation 1.114.

$$\gamma_{g,np} = j\beta_{g,np} = j\left(k_0^2 - k_c^2\right)^{1/2} = j\left[\left(\frac{2\pi}{\lambda_0}\right)^2 - \left(\frac{X'_{np}}{a}\right)^2\right]^{1/2} \tag{1.114}$$

For $k_0 > k_{c,nm}$, β is real and the wave will propagate. For $k_0 < k_{c,nm}$, β is imaginary and the wave will decay rapidly. Frequencies that define propagating and decaying waves are called cutoff frequencies. We may calculate cutoff frequencies by using Equation 1.115.

$$f_{c,nm} = \frac{cX'_{np}}{2\pi a} \tag{1.115}$$

We get the field components by solving Equations 1.110 and 1.111. Field components are written in Equation 1.116.

$$H_z = H_{0z}\left(J_n\left(\frac{X'_{np}r}{a}\right)\right)(\cos n\phi)e^{-j\beta_g z}$$

$$H_\phi = \frac{E_{0r}}{Z_g}\left(J_n\left(\frac{X'_{np}r}{a}\right)\right)(\sin n\phi)e^{-j\beta_g z}$$

$$H_r = \frac{E_{0\phi}}{Z_g}\left(J'_n\left(\frac{X'_{np}r}{a}\right)\right)(\cos n\phi)e^{-j\beta_g z} \tag{1.116}$$

$$E_\phi = E_{0\phi}\left(J'_n\left(\frac{X'_{np}r}{a}\right)\right)(\cos\phi)e^{-j\beta_g z}$$

$$E_r = E_{0r}\left(J_n\left(\frac{X'_{np}r}{a}\right)\right)(\sin\phi)e^{-j\beta_g z}$$

The impedance of the *np* modes is written in Equation 1.117.

$$Z_{g,np} = \frac{E_r}{H_\phi} = \frac{\omega\mu}{\beta_{g,np}} = \frac{\eta}{\sqrt{1 - \left(\dfrac{f_c}{f}\right)^2}} \tag{1.117}$$

1.9.2 TM WAVES IN A CIRCULAR WAVEGUIDE

In TM waves $h_z = 0$. e_z is given as the solution of Equation 1.118. The solution to Equation 1.118 is written in Equation 1.119.

$$\nabla^2 E_z = \gamma^2 E_z \tag{1.118}$$

$$E_z = E_{0z}(J_n(k_c r))(\cos n\phi)e^{\pm j\beta_g z} \tag{1.119}$$

The electric and magnetic fields are solutions of Maxwell equations as written in Equations 1.120 and 1.121.

$$\nabla X E = -j\omega\mu H \tag{1.120}$$

$$\nabla \times H = j\omega\varepsilon E \tag{1.121}$$

Field variation in the z direction may be written as $e^{-j\beta z}$. The derivative of this expression in the z direction may be written as $-j\beta e^{j\beta z}$. The electric and magnetic field components are solutions of Equations 1.122 and 1.123.

$$H_r = \frac{j\omega\varepsilon}{k_c^2}\frac{1}{r}\left(\frac{\partial H_z}{\partial \phi}\right)$$
$$H_\phi = \frac{-j\omega\varepsilon}{k_c^2}\left(\frac{\partial H_z}{\partial r}\right) \tag{1.122}$$

$$E_\phi = -\frac{-j\beta_g}{k_c^2}\frac{1}{r}\left(\frac{\partial E_z}{\partial \phi}\right)$$
$$E_r = \frac{-j\beta_g}{k_c^2}\left(\frac{\partial E_z}{\partial r}\right) \tag{1.123}$$

The E_r boundary condition is written in Equation 1.124.

$$E_z = 0 \quad @ \quad r = a \tag{1.124}$$

By applying the boundary conditions to Equation 1.119 we get the relations written in Equation 1.125. The solutions of Equation 1.125 are listed in Table 1.11.

$$E_z \big|(r = a) = H_{0z}(J_n(k_c a))(\cos n\phi)e^{-j\beta_g z} = 0$$
$$J_n(k_c a) = 0 \tag{1.125}$$

The wave number at cutoff is $k_{c,np}$. $k_{c,np}$ depends on the waveguide dimensions. The propagation constant $\gamma_{g,np}$ is given in Equation 1.126.

$$\gamma_{g,np} = j\beta_{g,np} = j\left(k_0^2 - k_c^2\right)^{1/2}$$
$$= j\left[\left(\frac{2\pi}{\lambda_0}\right)^2 - \left(\frac{X_{np}}{a}\right)^2\right]^{1/2} \tag{1.126}$$

For $k_0 > k_{c,nm}$, β is real and the wave will propagate. For $k_0 < k_{c,nm}$, β is imaginary and the wave will decay rapidly. Frequencies that define propagating and decaying waves are called cutoff frequencies. We may calculate cutoff frequencies by using Equation 1.127.

$$f_{c,nm} = \frac{cX_{np}}{2\pi a} \tag{1.127}$$

TABLE 1.11
Circular Waveguide T_m Modes

P	X_{np}, $n = 0$	X_{np}, $n = 1$	X_{np}, $n = 2$	X_{np}, $n = 3$	X_{np}, $n = 4$	X_{np}, $n = 5$
1	2.405	3.832	5.136	6.38	7.588	8.771
2	5.52	7.106	8.417	9.761	11.065	12.339
3	8.645	10.173	11.62	13.015	14.372	
4	11.792	13.324	14.796			

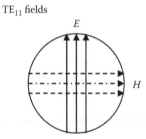

FIGURE 1.14   TE$_{11}$ fields in a circular waveguide.

We get the field components by solving Equations 1.122 and 1.123. Field components are written in Equation 1.128.

$$E_z = E_{0z}\left(J_n\left(\frac{X_{np}r}{a}\right)\right)(\cos n\phi)e^{-j\beta_g z}$$

$$H_\phi = \frac{E_{0r}}{Z_g}\left(J_n'\left(\frac{X_{np}r}{a}\right)\right)(\cos n\phi)e^{-j\beta_g z}$$

$$H_r = \frac{E_{0\phi}}{Z_g}\left(J_n\left(\frac{X_{np}r}{a}\right)\right)(\sin n\phi)e^{-j\beta_g z} \qquad (1.128)$$

$$E_\phi = E_{0\phi}\left(J_n\left(\frac{X_{np}r}{a}\right)\right)(\sin \phi)e^{-j\beta_g z}$$

$$E_r = E_{0r}\left(J_n'\left(\frac{X_{np}r}{a}\right)\right)(\cos \phi)e^{-j\beta_g z}$$

The impedance of the np modes is written in Equation 1.129.

$$Z_{g,np} = \frac{E_r}{H_\phi} = \frac{\beta_{g,np}}{\omega\varepsilon} = \eta\sqrt{1-\left(\frac{f_c}{f}\right)^2} \qquad (1.129)$$

The mode with the lowest cutoff frequency in circular waveguide is TE$_{11}$. TE$_{11}$ fields in a circular waveguide are shown in Figure 1.14.

REFERENCES

1. Ramo, S., Whinnery, J. R., and Van Duzer, T. *Fields and Waves in Communication Electronics*, 3rd ed. New York: John Wiley & Sons, 1994.
2. Collin, R. E. *Foundations for Microwave Engineering*. New York: McGraw-Hill, 1996.
3. Balanis, C. A. *Antenna Theory: Analysis and Design*, 2nd ed. New York: John Wiley & Sons, 1996.

4. Godara, L. C. (Ed.). *Handbook of Antennas in Wireless Communications*. Boca Raton, FL: CRC Press, 2002.

5. Kraus, J. D., and Marhefka, R. J. *Antennas for All Applications*, 3rd ed. New York: McGraw-Hill, 2002.

6. Ulaby, F. T. *Electromagnetics for Engineers*. Upper Saddle River, NJ: Pearson Education, 2004.

7. James, J. R., Hall, P. S., and Wood, C. *Microstrip Antenna Theory and Design*. The Institution of Engineering and Technology, 1981.

8. Sabban, A. *RF Engineering, Microwave and Antennas*. Tel Aviv, Israel: Saar Publication, 2014.

2 Basic Antenna Theory

2.1 INTRODUCTION TO ANTENNAS

Antennas are used to radiate efficiently electromagnetic energy in desired directions. Antennas match radio frequency systems, sources of electromagnetic energy, to space. All antennas may be used to receive or radiate energy. Antennas transmit or receive electromagnetic waves at radio frequencies. They convert electromagnetic radiation into electric current, or vice versa. Antennas are a necessary part of all communication links and radio equipment. They are used in systems such as radio and television broadcasting, point-to-point radio communication, wireless local area network (LAN), cell phones, radar, medical systems, and spacecraft communication. Antennas are most commonly employed in air or outer space, but can also be operated under water, on and inside the human body, or even through soil and rock at low frequencies for short distances. Physically, an antenna is an arrangement of one or more conductors. In transmitting mode, an alternating current is created in the elements by applying a voltage at the antenna terminals, causing the elements to radiate an electromagnetic field. In receiving mode, an electromagnetic field from another source induces an alternating current in the elements and a corresponding voltage at the antenna's terminals. Some receiving antennas (such as parabolic and horn) incorporate shaped reflective surfaces to receive the radio waves striking them and direct or focus them onto the actual conductive elements [1–11].

2.2 ANTENNA PARAMETERS

Antenna effective area (A_{eff}): The antenna area that contributes to the antenna directivity.

$$D = \frac{4\pi A_{eff}}{\lambda^2} \tag{2.1}$$

Antenna gain (G): The ratio between the amounts of energy propagating in a certain direction compared to the energy that would be propagating in the same direction if the antenna were not directional (isotropic radiator) is known as its gain.

Antenna impedance: Antenna impedance is the ratio of voltage at any given point along the antenna to the current at that point. Antenna impedance depends on the height of the antenna above the ground and the influence of surrounding objects. The impedance of a quarter wave monopole near a perfect ground is approximately 36 ohms. The impedance of a half wave dipole is approximately 75 ohms.

Azimuth (AZ): The angle from left to right from a reference point, from $0°$ to $360°$.

Beam width: The beam width is the angular range of the antenna pattern in which at least half of the maximum power is emitted. This angular range, of the major lobe, is defined as the points at which the field strength falls around 3 dB with respect to the maximum field strength.

Bore sight: The direction in space to which the antenna radiates maximum electromagnetic energy.

Directivity: The ratio between the amounts of energy propagating in a certain direction compared to the average energy radiated to all directions over a sphere.

$$D = \frac{P(\theta,\phi)\text{maximal}}{P(\theta,\phi)\text{average}} = 4\pi \frac{P(\theta,\phi)\text{maximal}}{P_{\text{rad}}} \qquad (2.2)$$

$$P(\theta,\phi)\text{average} = \frac{1}{4\pi}\iint P(\theta,\phi)\sin\theta \, d\theta \, d\phi = \frac{P_{\text{rad}}}{4\pi} \qquad (2.3)$$

$$D \sim \frac{4\pi}{\theta E \times \theta H} \qquad (2.4)$$

where:
θE = beam width in radian in *EL* plane
θH = beam width in radian in *AZ* plane

Elevation (EL): The *EL* angle is the angle from the horizontal (x, y) plane, from $-90°$ (down) to $+90°$ (up).

Isotropic radiator: Theoretical lossless radiator that radiates, or receives, equal electromagnetic energy in free space to all directions.

Main beam: The main beam is the region around the direction of maximum radiation, usually the region that is within 3 dB of the peak of the main lobe.

Phased arrays: Phased array antennas are electrically steerable. The physical antenna can be stationary. Phased arrays (smart antennas) incorporate active components for beam steering.

Radiated power: Radiated power is the total radiated power when the antenna is excited by a current or voltage of known intensity.

Radiation efficiency (α): The radiation efficiency is the ratio of power radiated to the total input power. The efficiency of an antenna takes into account losses, and is equal to the total radiated power divided by the radiated power of an ideal lossless antenna.

$$G = \alpha D \qquad (2.5)$$

$$\text{For small antennas} \left(l < \frac{\lambda}{2}\right) \quad G \cong \frac{41,000}{\theta E° \times \theta H°} \qquad (2.6)$$

For medium size antennas $(\lambda < l < 8\lambda)$ $G \cong \dfrac{31,000}{\theta E° \times \theta H°}$ (2.7)

For big antennas $(8\lambda < l)$ $G \cong \dfrac{27,000}{\theta E° \times \theta H°}$ (2.8)

A comparison of directivity and gain values for several antennas is given in Table 2.1.

Radiation pattern: A radiation pattern is the antenna-radiated field as a function of the direction in space. It is a way of plotting the radiated power from an antenna. This power is measured at various angles at a constant distance from the antenna.

Radiator: This is the basic element of an antenna. An antenna can be made up of multiple radiators.

Range: Antenna range is the radial range from an antenna to an object in space.

Side lobes: Side lobes are smaller beams that are away from the main beam. Side lobes present radiation in undesired directions. The side-lobe level is a parameter used to characterize the antenna radiation pattern. It is the maximum value of the side lobes away from the main beam and usually is expressed in decibels.

Steerable antennas:
- Arrays with switchable elements and partially mechanically and electronically steerable arrays.
- Hybrid antenna systems to fully electronically steerable arrays. Such systems can be equipped with phase and amplitude shifters for each element, or the design can be based on digital beam forming (DBF).
- This technique, in which the steering is performed directly on a digital level, allows the most flexible and powerful control of the antenna beam.

TABLE 2.1
Antenna Directivity versus Antenna Gain

Antenna Type	Directivity (dBi)	Gain (dBi)
Isotropic radiator	0	0
Dipole $\lambda/2$	2	2
Dipole above ground plane	6–4	6–4
Microstrip antenna	7–8	6–7
Yagi antenna	6–18	5–16
Helix antenna	7–20	6–18
Horn antenna	10–30	9–29
Reflector antenna	15–60	14–58

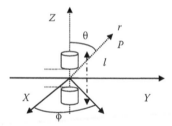

FIGURE 2.1   Dipole antenna.

2.3   DIPOLE ANTENNA

A dipole antenna is a small wire antenna. A dipole antenna consists of two straight conductors excited by a voltage fed via a transmission line as shown in Figure 2.1. Each side of the transmission line is connected to one of the conductors. The most common dipole is the half-wave dipole, in which each of the two conductors is approximately a quarter wavelength long, so the length of the antenna is a half-wavelength.

We can calculate the fields radiated from the dipole by using the potential function. The electric potential function is ϕ_l. The electric potential function is A. The potential function is given in Equation 2.9.

$$\phi_l = \frac{1}{4\pi\varepsilon_0} \int_c \frac{\rho_l e^{j(\omega t - \beta R)}}{R} dl$$

$$A_l = \frac{\mu_0}{4\pi} \int_c \frac{i e^{j(\omega t - \beta R)}}{R} dl$$

(2.9)

2.3.1   RADIATION FROM A SMALL DIPOLE

The length of a small dipole is small compared to wavelength and is called the elementary dipole. We may assume that the current along the elementary dipole is uniform. We can solve the wave equation in spherical coordinates by using the potential function given in Equation 2.9. The electromagnetic fields in a point $P(r, \theta, \varphi)$ is given in Equation 2.10. The electromagnetic fields in Equation 2.3 vary as $\frac{1}{r}, \frac{1}{r^2}, \frac{1}{r^3}$. For $r \ll 1$, the dominant component of the field varies as $\frac{1}{(r)^3}$ and is given in Equation 2.11. These fields are the dipole near fields. In this case the waves are standing waves and the energy oscillates in the antenna near zone and is not radiated to the open space. The real part of the Poynting vector is equal to zero. For $r \gg 1$, the dominant component of the field varies as $1/r$ as given in Equation 2.12. These fields are the dipole far fields.

$$E_r = \eta_0 \frac{ll_0 \cos\theta}{2\pi r^2}\left(1 - \frac{j}{\beta r}\right)e^{j(\omega t - \beta r)}$$

$$E_\theta = j\eta_0 \frac{\beta ll_0 \sin\theta}{4\pi r}\left(1 - \frac{j}{\beta r} - \frac{1}{(\beta r)^2}\right)e^{j(\omega t - \beta r)}$$

$$H_\phi = j\frac{\beta ll_0 \sin\theta}{4\pi r}\left(1 - \frac{j}{\beta r}\right)e^{j(\omega t - \beta r)} \qquad (2.10)$$

$$H_r = 0 \qquad H_\theta = 0 \qquad E_\phi = 0$$

$$I = I_0 \cos\omega t$$

$$E_r = -j\eta_0 \frac{ll_0 \cos\theta}{2\pi\beta r^3}e^{j(\omega t - \beta r)}$$

$$E_\theta = -j\eta_0 \frac{ll_0 \sin\theta}{4\pi\beta r^3}e^{j(\omega t - \beta r)} \qquad (2.11)$$

$$H_\phi = \frac{ll_0 \sin\theta}{4\pi r^2}e^{j(\omega t - \beta r)}$$

$$E_r = 0$$

$$E_\theta = j\eta_0 \frac{l\beta I_0 \sin\theta}{4\pi r}e^{j(\omega t - \beta r)} \qquad (2.12)$$

$$H_\phi = j\frac{l\beta I_0 \sin\theta}{4\pi r}e^{j(\omega t - \beta r)}$$

$$\frac{E_\theta}{H_\phi} = \eta_0 = \sqrt{\frac{\mu_0}{\varepsilon_0}} \qquad (2.13)$$

In the far fields the electromagnetic fields vary as $\frac{1}{r}$ and $\sin\theta$. Wave impedance in free space is given in Equation 2.13.

2.3.2 DIPOLE RADIATION PATTERN

The antenna radiation pattern represents the radiated fields in space at a point $P(r, \theta, \varphi)$ as function of θ, φ. The antenna radiation pattern is three dimensional. When φ is constant and θ varies we get the E-plane radiation pattern. When φ varies and θ is constant, usually $\theta = \pi/2$, we get the H-plane radiation pattern.

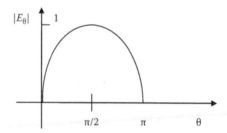

FIGURE 2.2   Dipole *E*-plane radiation pattern.

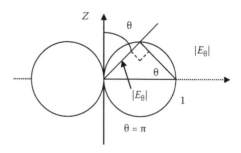

FIGURE 2.3   Dipole *E*-plane radiation pattern in a spherical coordinate system.

2.3.3   DIPOLE *E*-PLANE RADIATION PATTERN

The dipole *E*-plane radiation pattern is given in Equation 2.14 and presented in Figure 2.2.

$$|E_\theta| = \eta_0 \frac{l\beta I_0 |\sin\theta|}{4\pi r} \tag{2.14}$$

At a given point $P(r, \theta, \varphi)$ the dipole *E*-plane radiation pattern is given in Equation 2.15.

$$|E_\theta| = \eta_0 \frac{l\beta I_0 |\sin\theta|}{4\pi r} = A|\sin\theta|$$

$$\text{Choose} \quad A = 1 \tag{2.15}$$

$$|E_\theta| = |\sin\theta|$$

The dipole *E*-plane radiation pattern in spherical coordinate system is shown in Figure 2.3.

2.3.4   DIPOLE *H*-PLANE RADIATION PATTERN

For $\theta = \pi/2$ the dipole *H*-plane radiation pattern is given in Equation 2.10 and presented in Figure 2.4.

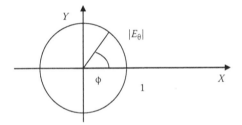

FIGURE 2.4   Dipole *H*-plane radiation pattern for $\theta = \pi/2$.

$$\left|E_\theta\right| = \eta_0 \frac{l\beta I_0}{4\pi r} \tag{2.16}$$

At a given point $P(r, \theta, \varphi)$ the dipole *H*-plane radiation pattern is given in Equation 2.17.

$$\left|E_\theta\right| = \eta_0 \frac{l\beta I_0 \left|\sin\theta\right|}{4\pi r} = A$$

$$\text{Choose} \quad A = 1 \tag{2.17}$$

$$\left|E_\theta\right| = 1$$

The dipole *H*-plane radiation pattern in the *x*–*y* plane is a circle with $r = 1$.

The radiation pattern of a vertical dipole is omnidirectional. It radiates equal power in all azimuthal directions perpendicular to the axis of the antenna. The dipole *H*-plane radiation pattern in spherical coordinate system is shown in Figure 2.4.

2.3.5   ANTENNA RADIATION PATTERN

A typical antenna radiation pattern is shown in Figure 2.5. The antenna main beam is measured between the points that the maximum relative field intensity *E* decays to 0.707*E*. Half of the radiated power is concentrated in the antenna main beam. The

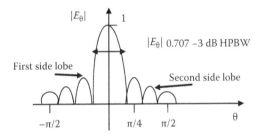

FIGURE 2.5   Antenna typical radiation pattern.

antenna main beam is called the 3 dB beam width. Radiation to an undesired direction is concentrated in the antenna side lobes.

For a dipole the power intensity varies as $(\sin^2 \theta)$. At $\theta = 45°$ and $\theta = 135°$ the radiated power equals half the power radiated toward $\theta = 90°$. The dipole beam width is

$$\theta = (135 - 45) = 90°.$$

2.3.6 DIPOLE DIRECTIVITY

Directivity is defined as the ratio between the amounts of energy propagating in a certain direction compared to the average energy radiated to all directions over a sphere as written in Equations 2.18 and 2.19.

$$D = \frac{P(\theta,\phi)\text{maximal}}{P(\theta,\phi)\text{average}} = 4\pi \frac{P(\theta,\phi)\text{maximal}}{P_{\text{rad}}} \qquad (2.18)$$

$$P(\theta,\phi)\text{average} = \frac{1}{4\pi} \iint P(\theta,\phi)\sin\theta \, d\theta \, d\phi = \frac{P_{\text{rad}}}{4\pi} \qquad (2.19)$$

The radiated power from a dipole is calculated by computing the Poynting vector **P** as given in Equation 2.20.

$$\mathbf{P} = 0.5(E \times H^*) = \frac{15\pi I_0^2 l^2 \sin^2\theta}{r^2\lambda^2}$$

$$W_T = \int_s \mathbf{P} \cdot ds = \frac{15\pi I_0^2 l^2}{\lambda^2} \int_0^\pi \sin^3\theta \, d\theta \int_0^{2\pi} d\phi = \frac{40\pi^2 I_0^2 l^2}{\lambda^2} \qquad (2.20)$$

The overall radiated energy is W_T. W_T is computed by integration of the power flow over an imaginary sphere surrounding the dipole. The power flow of an isotropic radiator is equal to W_T divided by the surrounding area of the sphere, $4\pi r^2$, as given in Equation 2.21. The dipole directivity at $\theta = 90°$ is 1.5 or 1.76 dB as shown in Equation 2.22.

$$\oint_s ds = r^2 \int_0^\pi \sin\theta \, d\theta \int_0^{2\pi} d\phi = 4\pi r^2$$

$$P_{\text{iso}} = \frac{W_T}{4\pi r^2} = \frac{10\pi I_0^2 l^2}{r^2\lambda^2} \qquad (2.21)$$

$$D = \frac{P}{P_{\text{iso}}} = 1.5\sin^2\theta$$

$$G_{dB} = 10 \log_{10} G = 10 \log_{10} 1.5 = 1.76 \text{ dB} \qquad (2.22)$$

For small antennas or for antennas without losses, $D = G$, losses are negligible. For a given θ and φ for small antennas the approximate directivity is given by Equation 2.23.

$$D = \frac{41,253}{\theta_{3 \, dB} \varphi_{3 \, dB}} \qquad (2.23)$$

$$G = \xi D \quad \xi = \text{Efficiency}$$

Antenna losses degrade the antenna efficiency. Antenna losses consist of conductor loss, dielectric loss, radiation loss, and mismatch losses. For resonant small antennas $\xi = 1$. For reflector and horn antennas the efficiency varies between $\xi = 0.5$ and $\xi = 0.7$.

2.3.7 ANTENNA IMPEDANCE

Antenna impedance determines the efficiency of transmitting and receiving energy in antennas. The dipole impedance is given in Equation 2.24.

$$R_{rad} = \frac{2W_T}{I_0^2}$$

For a dipole $\qquad (2.24)$

$$R_{rad} = \frac{80\pi^2 l^2}{\lambda^2}$$

2.3.8 IMPEDANCE OF A FOLDED DIPOLE

A folded dipole is a half-wave dipole with an additional wire connecting its two ends. If the additional wire has the same diameter and cross section as the dipole, two nearly identical radiating currents are generated. The resulting far-field emission pattern is nearly identical to the one for the single-wire dipole described previously, but at resonance its feed point impedance R_{rad-f} is four times the radiation resistance of a dipole. This is because for a fixed amount of power, the total radiating current I_0 is equal to twice the current in each wire and thus equal to twice the current at the feed point. Equating the average radiated power to the average power delivered at the feed point, we obtain that $R_{rad-f} = 4\,R_{rad} = 300\ \Omega$. The folded dipole has a wider bandwidth than a single dipole.

2.4 BASIC APERTURE ANTENNAS

Reflector and horn antennas are defined as aperture antennas. The longest dimension of an aperture antenna is greater than several wavelengths.

2.4.1 THE PARABOLIC REFLECTOR ANTENNA

The parabolic reflector antenna [3] consists of a radiating feed that is used to illuminate a reflector that is curved in the form of an accurate parabolic with diameter D as presented in Figure 2.6. This shape enables evert beam to be obtained. To provide the optimum illumination of the reflecting surface, the level of the parabola illumination should be greater by 10 dB in the center than that at the parabola edges. The parabolic reflector antenna gain may be calculated by using Equation 2.25. α is the parabolic reflector antenna efficiency.

Parabolic reflector antenna gain:

$$G \cong 10 \log_{10}\left(\alpha \frac{(\pi D)^2}{\lambda^2}\right) \tag{2.25}$$

Reflector geometry is presented in Figure 2.7.

The following relations given in Equations 2.26 to 2.29 can be derived from the reflector geometry.

$$PQ = r' \cos \theta' \tag{2.26}$$

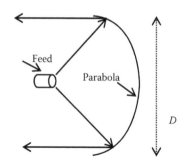

FIGURE 2.6 Parabolic antenna.

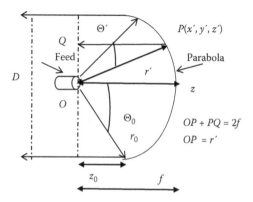

FIGURE 2.7 Reflector geometry.

$$2f = r'(1 + \cos \theta') \tag{2.27}$$

$$2f = r' + r'\cos \theta' = \sqrt{(x')^2 + (y')^2 + (z')^2} + z' \tag{2.28}$$

The relation between the reflector diameter D and θ is given in Equations 2.29 to 2.31.

$$\theta_0 = \tan^{-1} \frac{\dfrac{D}{2}}{z_0} \tag{2.29}$$

$$z_0 = f - \frac{\left(\dfrac{D}{2}\right)^2}{4f} \tag{2.30}$$

$$\theta_0 = \tan^{-1} \left| \frac{\dfrac{D}{2}}{z_0} \right| = \tan^{-1} \left| \frac{\dfrac{D}{2}}{f - \dfrac{\left(\dfrac{D}{2}\right)^2}{4f}} \right| = \tan^{-1} \left| \frac{\dfrac{f}{2D}}{\left(\dfrac{f}{D}\right)^2 - \dfrac{1}{16}} \right| \tag{2.31}$$

$$f = \frac{D}{4} \cot\left(\frac{\theta_0}{2}\right) \tag{2.32}$$

2.4.2 REFLECTOR DIRECTIVITY

Reflector directivity is a function of the reflector geometry and feed radiation characteristics as given in Equations 2.33 and 2.34.

$$D_0 = \frac{4\pi U_{max}}{P_{rad}} = \frac{16\pi^2}{\lambda^2} f^2 \left| \int_0^{\theta_0} \sqrt{G_F(\theta')} \tan\left(\frac{\theta'}{2}\right) d\theta' \right|^2 \tag{2.33}$$

$$D_0 = \frac{(\pi D)^2}{\lambda^2} \left[\cot^2\left(\frac{\theta_0}{2}\right) \left| \int_0^{\theta_0} \sqrt{G_F(\theta')} \tan\left(\frac{\theta'}{2}\right) d\theta' \right|^2 \right] \tag{2.34}$$

The reflector aperture efficiency is given in Equation 2.35. The feed radiation pattern may be presented as in Equation 2.36.

$$\epsilon_{ap} = \left[\cot^2\left(\frac{\theta_0}{2}\right) \left| \int_0^{\theta_0} \sqrt{G_F(\theta')} \tan\left(\frac{\theta'}{2}\right) d\theta' \right|^2 \right] \tag{2.35}$$

$$G_F(\theta') = G_0^n \cos^n(\theta') \quad \text{for} \quad 0 \le \theta' \le \frac{\pi}{2}$$

$$G_F(\theta') = 0 \quad \text{for} \quad \frac{\pi}{2} < \theta' \le \pi \tag{2.36}$$

$$\int_0^{\pi/2} G_0^n \cos^n(\theta') \sin(\theta') \, d\theta' = 2$$

where

$$G_0^n = 2(n+1)$$

Uniform illumination of the reflector aperture may be achieved if $G_F(\theta')$ is given by Equation 2.37.

$$G_F(\theta') = \sec^4\left(\frac{\theta'}{2}\right) \quad \text{for} \quad 0 \le \theta' \le \frac{\pi}{2}$$

$$G_F(\theta') = 0 \quad \text{for} \quad \frac{\pi}{2} < \theta' \le \pi \tag{2.37}$$

The reflector aperture efficiency is computed by multiplying all the antenna efficiencies due to spillover, blockage, taper, phase error, cross polarization losses, and random error over the reflector surface.

$$\epsilon_{ap} = \epsilon_s \epsilon_t \epsilon_b \epsilon_x \epsilon_p \epsilon_r$$

where

ϵ_s = spillover efficiency, written in Equation 2.38
ϵ_t = taper efficiency, written in Equation 2.39
ϵ_b = blockage efficiency
ϵ_p = phase efficiency
ϵ_x = cross polarization efficiency
ϵ_r = random error over the reflector surface efficiency

$$\epsilon_s = \frac{\displaystyle\int_0^{\theta_0} G_F(\theta') \sin\theta' \, d\theta'}{\displaystyle\int_\pi^{\theta_0} G_F(\theta') \sin\theta' \, d\theta'} \tag{2.38}$$

$$\epsilon_t = 2\cot^2\left(\frac{\theta_0}{2}\right) \frac{\left|\displaystyle\int_0^{\theta_0} \sqrt{G_F(\theta')} \tan\left(\frac{\theta'}{2}\right) d\theta'\right|^2}{\displaystyle\int_0^{\theta_0} G_F(\theta') \sin\theta' \, d\theta'} \tag{2.39}$$

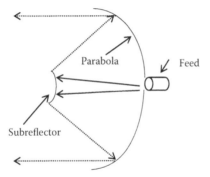

FIGURE 2.8 Cassegrain feed system.

In the literature [1] we can find graphs that present the reflector antenna efficiencies as a function of the reflector antenna geometry and feed radiation pattern. However, Equations 2.34 to 2.40 give us a good approximation of the reflector directivity.

$$D_0 = \frac{(\pi D)^2 \epsilon_{ap}}{\lambda^2} \qquad (2.40)$$

2.4.3 CASSEGRAIN REFLECTOR

The parabolic reflector or dish antenna consists of a radiating element that may be a simple dipole or a waveguide horn antenna. This is placed at the center of the metallic parabolic reflecting surface as shown in Figure 2.8. The energy from the radiating element is arranged so that it illuminates the subreflecting surface. The energy from the subreflector is arranged so that it illuminates the main reflecting surface. Once the energy is reflected it leaves the antenna system in a narrow beam.

2.5 HORN ANTENNAS

Horn antennas are used as a feed element for radio astronomy, satellite tracking and communication reflector antennas, and phased arrays radiating elements, used in antenna calibration and measurements. Figure 2.9a shows an E-plane sectoral horn. Figure 2.9b shows an H-plane sectoral horn. Figure 2.9c shows a pyramidal horn. Figure 2.9d shows a conical horn.

2.5.1 E-PLANE SECTORAL HORN

Figure 2.10 shows an E-plane sectoral horn. Horn antennas are fed by a waveguide. The excited mode is TE_{10}. Fields expressions over the horn aperture are similar to the fields of a TE_{10} mode in a rectangular waveguide with the aperture dimensions of a, b_1. The fields in the antenna aperture are given in Equations 2.41 to 2.43.

$$E'_y(x', y') \approx E_1 \cos\left(\frac{\pi}{a}x'\right) e^{-j\left[ky'^2/(2\rho_1)\right]} \qquad (2.41)$$

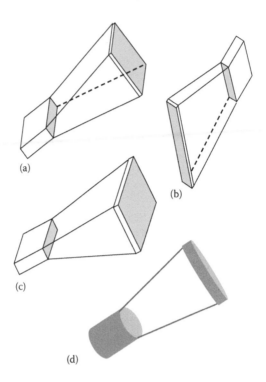

(a)

(b)

(c)

(d)

FIGURE 2.9 (a) *E*-plane sectoral horn. (b) *H*-plane sectoral horn. (c) Pyramidal horn. (d) Conical horn.

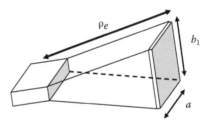

FIGURE 2.10 *E*-plane sectoral horn.

$$H'_x(x',y') \approx \frac{E_1}{\eta} \cos\left(\frac{\pi}{a}x'\right) e^{-j\left[ky'^2/(2\rho_1)\right]} \tag{2.42}$$

$$H'_z(x',y') \approx jE_1 \left(\frac{\pi}{ka\eta}\right) \sin\left(\frac{\pi}{a}x'\right) e^{-j\left[ky'^2/(2\rho_1)\right]} \tag{2.43}$$

The horn length is ρ_1, as shown in Figure 2.11. The extra distance along the aperture sides compared with the distance to the center is δ and is given by Equation 2.44.

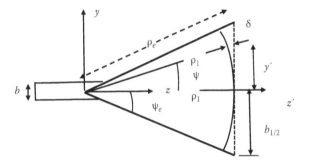

FIGURE 2.11 *E*-plane sectoral horn geometry.

$$\delta = \rho_e - \sqrt{\rho_e - \left(\frac{b_1}{2}\right)^2} = \rho_e \left(1 - \sqrt{1 - \left(\frac{b_1}{2\rho_e}\right)^2}\right) = \frac{b_1^2}{8\rho_e} \tag{2.44}$$

$$\frac{\delta}{\lambda} = S = \frac{b_1^2}{8\lambda\rho_e} \tag{2.45}$$

S represents the quadratic phase distribution as given in Equation 2.45.

$$(\delta(y') + \rho_1)^2 = \rho_1^2 + (y')^2 \tag{2.46}$$

$$\delta(y') = -\rho_1 + \sqrt{\rho_1^2 + (y')^2} = -\rho_1 + \rho_1\sqrt{1 + \left(\frac{y'}{\rho_1}\right)^2} \tag{2.47}$$

The maximum phase deviation at the aperture $\varnothing_{max}$ is given by Equation 2.48.

$$\varnothing_{max} = k\delta(y'):\left(y' = \frac{b_1}{2}\right) = \frac{kb_1^2}{8\rho_1} \tag{2.48}$$

The total flare angle of the horn, $2\psi_e$, is given in Equation 2.49.

$$2\psi_e = 2\tan^{-1}\left(\frac{b_1}{2\rho_1}\right) \tag{2.49}$$

Directivity of the *E*-Plane Horn

The maximum radiation is given by Equation 2.50.

$$U_{max} = \frac{r^2}{2\eta}|E|_{max}^2 \tag{2.50}$$

$$U_{max} = \frac{2ka^2\rho_1}{\pi^3\eta}|E|^2|F(t)|^2 \tag{2.51}$$

$$|F(t)|^2 = \left[C^2\left(\frac{b_1}{\sqrt{2\lambda\rho_1}}\right) + S^2\left(\frac{b_1}{\sqrt{2\lambda\rho_1}}\right)\right] \tag{2.52}$$

C and S are Fresnel integers and are given in Table 2.2. The total radiated power by the horn is given in Equation 2.53.

$$P_{rad} = \frac{ab_1}{4\eta}|E|^2 \tag{2.53}$$

The directivity of E-plane horn D_E is given in Equation 2.54.

$$D_E = \frac{4\pi U_{max}}{P_{rad}} = \frac{64a\rho_1}{\pi\lambda b_1}\left[C^2\left(\frac{b_1}{\sqrt{2\lambda\rho_1}}\right) + S^2\left(\frac{b_1}{\sqrt{2\lambda\rho_1}}\right)\right] \tag{2.54}$$

Figure 2.12 presents the H-plane horn radiation pattern as function of S,

where
$$S = \frac{b_1^2}{8\lambda\rho_e}.$$

2.5.2 H-PLANE SECTORAL HORN

An H-plane sectoral horn is shown in Figure 2.13. H-plane sectoral horn geometry is shown in Figure 2.14. Fields expressions over the horn aperture are similar to the fields of a TE_{10} mode in rectangular waveguide with the aperture dimensions of a, b_1. The fields in the antenna aperture are written in Equations 2.55 and 2.56.

$$E'_y(x', y') \approx E_2 \cos\left(\frac{\pi}{a}x'\right)e^{-j\left[kx'^2/(2\rho_2)\right]} \tag{2.55}$$

$$H'_x(x', y') \approx \frac{E_2}{\eta} \cos\left(\frac{\pi}{a}x'\right)e^{-j\left[kx'^2/(2\rho_2)\right]} \tag{2.56}$$

The horn length is ρ_l. The extra distance along the aperture sides compared with the distance to the center is δ and is given by Equation 2.57.

TABLE 2.2
Fresnel Integers

x	$C_1(x)$	$S_1(x)$	$C(x)$	$S(x)$
0.0	0.62666	0.62666	0.0	0.0
0.1	0.52666	0.62632	0.10000	0.00052
0.2	0.42669	0.62399	0.19992	0.00419
0.3	0.32690	0.61766	0.29940	0.01412
0.4	0.22768	0.60536	0.39748	0.03336
0.5	0.12977	0.58518	0.49234	0.06473
0.6	0.03439	0.55532	0.58110	0.11054
0.7	−0.05672	0.51427	0.65965	0.17214
0.8	−0.14119	0.46092	0.72284	0.24934
0.9	−0.21606	0.39481	0.76482	0.33978
1.0	−0.27787	0.31639	0.77989	0.43826
1.1	−0.32285	0.22728	0.76381	0.53650
1.2	−0.34729	0.13054	0.71544	0.62340
1.3	−0.34803	0.03081	0.63855	0.68633
1.4	−0.32312	−0.06573	0.54310	0.71353
1.5	−0.27253	−0.15158	0.44526	0.69751
1.6	−0.19886	−0.21861	0.36546	0.63889
1.7	−0.10790	−0.25905	0.32383	0.54920
1.8	−0.00871	−0.26682	0.33363	0.45094
1.9	0.08680	−0.23918	0.39447	0.37335
2.0	0.16520	−0.17812	0.48825	0.34342
2.1	0.21359	−0.09141	0.58156	0.37427
2.2	0.22242	0.00743	0.63629	0.45570
2.3	0.18833	0.10054	0.62656	0.55315
2.4	0.11650	0.16879	0.55496	0.61969
2.5	0.02135	0.19614	0.45742	0.61918
2.6	−0.07518	0.17454	0.38894	0.54999
2.7	−0.14816	0.10789	0.39249	0.45292
2.8	−0.17646	0.01329	0.46749	0.39153
2.9	−0.15021	−0.08181	0.56237	0.41014
3.0	−0.07621	−0.14690	0.60572	0.49631
3.1	0.02152	−0.15883	0.56160	0.58181
3.2	1.10791	−0.11181	0.46632	0.59335
3.3	0.14907	−0.02260	0.40570	0.51929
3.4	0.12691	0.07301	0.43849	0.42965
3.5	0.04965	0.13335	0.53257	0.41525
3.6	−0.04819	0.12973	0.58795	0.49231
3.7	−0.11929	0.06258	0.54195	0.57498
3.8	−0.12649	−0.03483	0.44810	0.56562
3.9	−0.06469	−0.11030	0.42233	0.47521
4.0	0.03219	−0.12048	0.49842	0.42052
4.1	0.10690	−0.05815	0.57369	0.47580

(Continued)

TABLE 2.2 (CONTINUED)
Fresnel Integers

x	$C_1(x)$	$S_1(x)$	$C(x)$	$S(x)$
4.2	0.11228	0.03885	0.54172	0.56320
4.3	0.04374	0.10751	0.44944	0.55400
4.4	−0.05287	0.10038	0.43833	0.46227
4.5	−0.10884	0.02149	0.52602	0.43427
4.6	−0.08188	−0.07126	0.56724	0.51619
4.7	0.00810	−0.10594	0.49143	0.56715
4.8	0.08905	−0.05381	0.43380	0.49675
4.9	0.09277	0.04224	0.50016	0.43507
5.0	0.01519	0.09874	0.56363	0.49919
5.1	−0.07411	0.06405	0.49979	0.56239
5.2	−0.09125	−0.03004	0.43889	0.49688
5.3	−0.01892	−0.09235	0.50778	0.44047
5.4	0.07063	−0.05976	0.55723	0.51403
5.5	0.08408	0.03440	0.47843	0.55369
5.6	0.00641	0.08900	0.45171	0.47004
5.7	−0.07642	0.04296	0.53846	0.45953
5.8	−0.06919	−0.05135	0.52984	0.54604
5.9	0.01998	−0.08231	0.44859	0.51633
6.0	0.08245	−0.01181	0.49953	0.44696
6.1	0.03946	0.07180	0.54950	0.51647
6.2	−0.05363	0.06018	0.46761	0.53982
6.3	−0.07284	−0.03144	0.47600	0.45555
6.4	0.00835	−0.07765	0.54960	0.49649
6.5	0.07574	−0.01326	0.48161	0.54538
6.6	0.03183	0.06872	0.46899	0.46307
6.7	−0.05828	0.04658	0.54674	0.49150
6.8	−0.05734	−0.04600	0.48307	0.54364
6.9	0.03317	−0.06440	0.47322	0.46244
7.0	0.06832	0.02077	0.54547	0.49970
7.1	−0.00944	0.06977	0.47332	0.53602
7.2	−0.06943	0.00041	0.48874	0.45725
7.3	−0.00864	−0.06793	0.53927	0.51894
7.4	0.06582	−0.01521	0.46010	0.51607
7.5	0.02018	0.06353	0.51601	0.46070
7.6	−0.06137	0.02367	0.51564	0.53885
7.7	−0.02580	−0.05958	0.46278	0.48202
7.8	0.05828	−0.02668	0.53947	0.48964
7.9	0.02638	0.05752	0.47598	0.53235
8.0	−0.05730	0.02494	0.49980	0.46021
8.1	−0.02238	−0.05752	0.52275	0.53204
8.2	0.05803	−0.01870	0.46384	0.48589
8.3	0.01387	0.05861	0.53775	0.49323

(Continued)

TABLE 2.2 (CONTINUED)

Fresnel Integers

x	$C_1(x)$	$S_1(x)$	$C(x)$	$S(x)$
8.4	−0.05899	0.00789	0.47092	0.52429
8.5	−0.00080	−0.05881	0.51417	0.46534
8.6	0.05767	0.00729	0.50249	0.53693
8.7	−0.01616	0.05515	0.48274	0.46774
8.8	−0.05079	−0.02545	0.52797	0.52294
8.9	0.03461	−0.04425	0.46612	0.48856
9.0	0.03526	0.04293	0.53537	0.49985
9.1	−0.04951	0.02381	0.46661	0.51042
9.2	−0.01021	−0.05338	0.52914	0.48135
9.3	0.05354	0.00485	0.47628	0.52467
9.4	−0.02020	0.04920	0.51803	0.47134
9.5	−0.03995	−0.03426	0.48729	0.53100
9.6	0.04513	−0.02599	0.50813	0.46786
9.7	0.00837	0.05086	0.49549	0.53250
9.8	−0.04983	−0.01094	0.50192	0.46758
9.9	0.02916	−0.04124	0.49961	0.53215
10.0	0.02554	0.04298	0.49989	0.46817
10.1	−0.04927	0.00478	0.49961	0.53151
10.2	0.01738	−0.04583	0.50186	0.46885
10.3	0.03233	0.03621	0.49575	0.53061
10.4	−0.04681	0.01094	0.50751	0.47033
10.5	0.01360	−0.04563	0.48849	0.52804

$$\delta = \rho_h - \sqrt{\rho_h - \left(\frac{a_1}{2}\right)^2} = \rho_h\left(1 - \sqrt{1 - \left(\frac{a_1}{2\rho_h}\right)^2}\right) = \frac{a_1^2}{8\rho_h} \tag{2.57}$$

$$\frac{\delta}{\lambda} = S = \frac{a_1^2}{8\lambda\rho_h} \tag{2.58}$$

S represents the quadratic phase distribution as written in Equation 2.58.

$$(\delta(x') + \rho_2)^2 = \rho_2^2 + (x')^2 \tag{2.59}$$

$$\delta(x') = -\rho_2 + \sqrt{\rho_2^2 + (x')^2} = -\rho_2 + \rho_1\sqrt{1 + \left(\frac{x'}{\rho_2}\right)^2} \tag{2.60}$$

The maximum phase deviation at the aperture $\varnothing_{max}$ is given by Equation 2.61.

$$\varnothing_{max} = k\delta(x'):\left(x' = \frac{a_1}{2}\right) = \frac{ka_1^2}{8\rho_2} \tag{2.61}$$

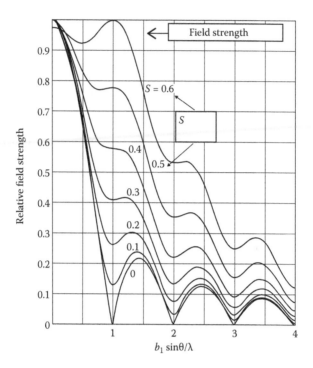

FIGURE 2.12 *H*-plane horn radiation pattern as function of S, $S = \dfrac{b_1^2}{8\lambda\rho_e}$.

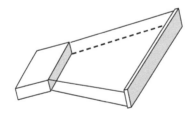

FIGURE 2.13 *H*-plane sectoral horn.

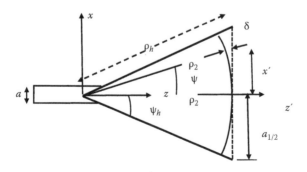

FIGURE 2.14 *E*-plane sectoral horn geometry.

The total flare angle of the horn, $2\psi_e$, is given in Equation 2.62.

$$2\psi_h = 2\tan^{-1}\left(\frac{a_1}{2\rho_2}\right) \tag{2.62}$$

Directivity of the *E*-Plane Horn

The maximum radiation is given by Equations 2.63 and 2.64.

$$U_{max} = \frac{r^2}{2\eta}|E|^2_{max} \tag{2.63}$$

$$U_{max} = \frac{b^2\rho_2}{4\lambda\eta}|E|^2|F(t)|^2 \tag{2.64}$$

$$|F(t)|^2 = [(C(u) - C(v))^2 + (S(u) - S(v))^2] \tag{2.65}$$

where $u = \dfrac{1}{\sqrt{2}}\left(\dfrac{\sqrt{\rho_2\lambda}}{a_1} - \dfrac{a_1}{\sqrt{\rho_2\lambda}}\right)$ and $v = \dfrac{1}{\sqrt{2}}\left(\dfrac{\sqrt{\rho_2\lambda}}{a_1} + \dfrac{a_1}{\sqrt{\rho_2\lambda}}\right)$

C and *S* are Fresnel integers. The total radiated power by the horn is given in Equation 2.66.

$$P_{rad} = \frac{ab_1}{4\eta}|E|^2 \tag{2.66}$$

The directivity of *H*-plane horn D_H is given in Equation 2.67.

$$D_H = \frac{4\pi U_{max}}{P_{rad}} = \frac{4b\pi\rho_2}{\lambda a_1}[(C(u) - C(v))^2 + (S(u) - S(v))]^2 \tag{2.67}$$

An *H*-plane horn radiation pattern as a function of *S*, $S = \dfrac{a_1^2}{8\lambda\rho_h}$, is shown in Figure 2.15.

2.5.3 PYRAMIDAL HORN ANTENNA

The pyramidal horn antenna is a combination of the *E* and *H* horns as shown in Figure 2.16. The pyramidal horn antenna is realizable only if $\rho_h = \rho_e$.

The directivity of pyramidal horn antenna D_P is given in Equation 2.68.

$$D_P = \frac{4\pi U_{max}}{P_{rad}} = \frac{\pi\lambda^2}{32ab}D_E D_H \tag{2.68}$$

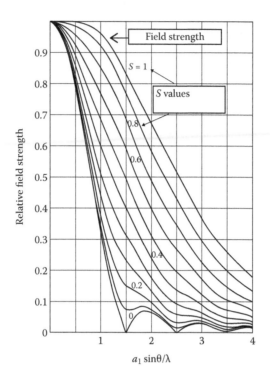

FIGURE 2.15 *H*-plane horn radiation pattern as a function of *S*, $S = \dfrac{a_1^2}{8\lambda\rho_h}$.

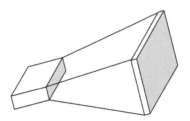

FIGURE 2.16 Pyramidal horn antenna.

Kraus [3] gives the following approximation for pyramidal horn beam width.

$$\theta_{3\,dB}^e = \frac{56}{A_{e\lambda}} \tag{2.69}$$

$$\theta_{3\,dB}^h = \frac{67}{A_{h\lambda}} \tag{2.70}$$

$$\theta^e_{10\,\text{dB}} = \frac{100.8}{A_{e\lambda}} \tag{2.71}$$

$$\theta^h_{10\,\text{dB}} = \frac{120.6}{A_{h\lambda}} \tag{2.72}$$

$A_{e\lambda}$ is the aperture dimensions in wavelength in the the E plane. $A_{h\lambda}$ is the aperture dimensions in wavelength in the H plane. The pyramidal horn gain is given by Equation 2.73.

$$G = 10\log_{10} 4.5 A_{h\lambda} A_{e\lambda} \text{ dBi} \tag{2.73}$$

The relative power at any angle, $P^h_{\text{dB}}(\theta)$, is given approximately by Equation 2.74.

$$P^h_{\text{dB}}(\theta) = 10\left(\frac{\theta}{\theta^h_{10\,\text{dB}}}\right)^2 \tag{2.74}$$

REFERENCES

1. Balanis, C. A. *Antenna Theory: Analysis and Design*, 2nd ed. New York: John Wiley & Sons, 1996
2. Godara, L. C. (Ed.). *Handbook of Antennas in Wireless Communications*. Boca Raton, FL: CRC Press, 2002.
3. Kraus, J. D., and Marhefka, R. J. *Antennas for All Applications*, 3rd ed. New York: McGraw-Hill, 2002.
4. James, J. R., Hall, P. S., and Wood, C. *Microstrip Antenna Theory and Design*. London: The Institution of Engineering and Technology, 1981.
5. Sabban, A., and Gupta, K. C. Characterization of Radiation Loss from Microstrip Discontinuities Using a Multiport Network Modeling Approach. *IEEE Transactions on Microwave Theory and Techniques*, 39(4): 705–712, 1991.
6. Sabban, A. *Multiport Network Model for Evaluating Radiation Loss and Coupling Among Discontinuities in Microstrip Circuits*. PhD thesis, University of Colorado at Boulder, January 1991.
7. Kathei, P. B., and Alexopoulos, N. G. Frequency-Dependent Characteristic of Microstrip, Discontinuities in Millimeter-Wave Integrated Circuits. *IEEE Transactions on Microwave Theory and Techniques*, 33: 1029–1035, 1985.
8. Sabban, A. A New Wideband Stacked Microstrip Antenna. In *IEEE Antenna and Propagation Symposium*, Houston, TX, June 1983.
9. Sabban, A., and Navon, E. A MM-Waves Microstrip Antenna Array. In *IEEE Symposium*, Tel Aviv, Israel, March 1983.
10. Sabban, A. Wideband Microstrip Antenna Arrays. In *IEEE Antenna and Propagation Symposium MELCOM*, Tel Aviv, Israel, June 1981.
11. Sabban, A. *RF Engineering, Microwave and Antennas*. Tel Aviv, Israel: Saar Publication, 2014.

3 Low-Visibility Printed Antennas

3.1 MICROSTRIP ANTENNAS

Microstrip antennas possess attractive features such as low profile, flexible, light weight, small volume, and low production cost. Microstrip antennas have been widely presented in books and papers in the last decade [1–18]. Microstrip antennas may be employed in communication links, seekers, and in biomedical systems.

3.1.1 INTRODUCTION TO MICROSTRIP ANTENNAS

Microstrip antennas are printed on a dielectric substrate with low dielectric losses. A cross section of the microstrip antenna is shown in Figure 3.1. Microstrip antennas are thin patches etched on a dielectric substrate ε_r. The substrate thickness, H, is less than 0.1 λ.

Advantages of microstrip antennas:

- Low cost to fabricate
- Conformal structures are possible
- Easy to form a large uniform array with half-wavelength spacing
- Light weight and low volume

Disadvantages of microstrip antennas:

- Limited bandwidth (usually 1%–5%, but much more is possible with increased complexity)
- Low power handling

The electric field along the radiating edges is shown in Figure 3.2. The magnetic field is perpendicular to the E-field according to Maxwell's equations. At the edge of the strip ($X/L = 0$ and $X/L = 1$) the H-field drops to zero; because there is no conductor to carry the radio frequency (RF) current, it is maximum in the center. The E-field intensity is at maximum magnitude (and opposite polarity) at the edges ($X/L = 0$ and $X/L = 1$) and zero at the center. The ratio of the E- to H-field is proportional to the impedance that we see when we feed the patch. Microstrip antennas may be fed by a microstrip line or by a coaxial line or probe feed. By adjusting the location of the feed point between the center and the edge, we can get any impedance, including 50 Ω. Microstrip antenna shape may be square, rectangular, triangle, circle, or any arbitrary shape as shown in Figure 3.3.

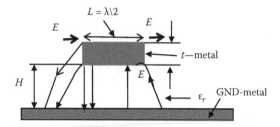

FIGURE 3.1 Microstrip antenna cross section.

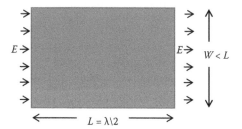

FIGURE 3.2 Rectangular microstrip antenna.

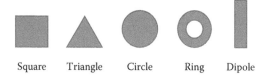

Square   Triangle   Circle   Ring   Dipole

FIGURE 3.3 Microstrip antenna shapes.

The dielectric constant that controls the resonance of the antenna is the effective dielectric constant of the microstrip line. The antenna dimension W is given by Equation 3.1.

$$W = \frac{c}{2f\sqrt{\epsilon_{\text{eff}}}} \tag{3.1}$$

The antenna bandwidth is given in Equation 3.2.

$$BW = \frac{H}{\sqrt{\epsilon_{\text{eff}}}} \tag{3.2}$$

The gain of microstrip antenna is between 0 dBi and 7 dBi. The microstrip antenna gain is a function of the antenna dimensions and configuration. We may

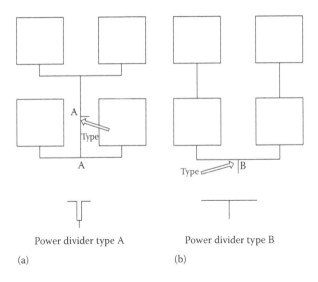

Power divider type A   Power divider type B

(a)   (b)

FIGURE 3.4   Configuration of microstrip antenna array. (a) Parallel feed network. (b) Parallel series feed network.

increase printed antenna gain by using a microstrip antenna array configuration. In a microstrip antenna array the benefit of a compact low-cost feed network is attained by integrating the RF feed network with the radiating elements on the same substrate. Microstrip antenna feed networks are presented in Figure 3.4. Figure 3.4a shows a parallel feed network. Figure 3.4b shows a parallel series feed network.

3.1.2   TRANSMISSION LINE MODEL OF MICROSTRIP ANTENNAS

In the transmission line model (TLM) of patch microstrip antennas the antenna is represented as two slots connected by a transmission line (Figure 3.5). TLM is not an accurate model. However, it gives a good physical understanding of patch

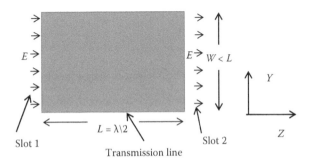

FIGURE 3.5   Transmission line model of patch microstrip antennas.

microstrip antennas. The electric fields along and underneath the patch depend on the z coordinate.

$$E_x \sim \cos\left(\frac{\pi z}{L_{\text{eff}}}\right) \qquad (3.3)$$

At $z = 0$ and $z = L_{\text{eff}}$ the electric field is maximum. At $z = \dfrac{L_{\text{eff}}}{2}$ the electric field equals zero.

For $\dfrac{H}{\lambda_0} < 0.1$ the electric field distribution along the x-axis is assumed to be uniform. The slot admittance may be written as given in Equations 3.4 and 3.5.

$$G = \frac{W}{120\lambda_0}\left[1 - \frac{1}{24}\left(\frac{2\pi H}{\lambda_0}\right)^2\right] \quad \text{for} \quad \frac{H}{\lambda_0} < 0.1 \qquad (3.4)$$

$$B = \frac{W}{120\lambda_0}\left[1 - 0.636 \ln\left(\frac{2\pi H}{\lambda_0}\right)^2\right] \quad \text{for} \quad \frac{H}{\lambda_0} < 0.1 \qquad (3.5)$$

B represents the capacitive nature of the slot. $G = 1/R$, where R represents the radiation losses. When the antenna is resonant the susceptances of both slots cancel out at the feed point for any position of the feed point along the patch. However, the patch admittance depends on the feed point position along the z-axis. At the feed point the slot admittance is transformed by the equivalent length of the transmission line. The width W of the microstrip antenna controls the input impedance. Larger widths can increase the bandwidth. For a square patch antenna fed by a microstrip line, the input impedance is around 300 Ω. By increasing the width, the impedance can be reduced.

$$Y(l_1) = Z_0 \frac{1 + j\dfrac{Z_L}{Z_0}\tan\beta l_1}{\dfrac{Z_L}{Z_0} + j\tan\beta l_1} = Y_1 \qquad (3.6)$$

$$Y_{in} = Y_1 + Y_2$$

3.1.3 HIGHER-ORDER TRANSMISSION MODES IN MICROSTRIP ANTENNAS

To prevent higher-order transmission modes we should limit the thickness of the microstrip substrate to 10% of a wavelength. The cutoff frequency of the higher-order mode is given in Equation 3.7.

$$f_c = \frac{c}{4H\sqrt{\varepsilon - 1}} \qquad (3.7)$$

3.1.4 EFFECTIVE DIELECTRIC CONSTANT

A part of the fields in the microstrip antenna structure exists in air and the other part of the fields exists in the dielectric substrate. The effective dielectric constant is somewhat less than the substrate's dielectric constant. The effective dielectric constant of the microstrip line may be calculated by Equations 3.8 and 3.9 as a function of W/H.

For $\left(\dfrac{W}{H}\right) < 1$

$$\varepsilon_e = \frac{\varepsilon_r + 1}{2} + \frac{\varepsilon_r - 1}{2}\left[\left(1 + 12\left(\frac{H}{W}\right)\right)^{-0.5} + 0.04\left(1 - \left(\frac{W}{H}\right)\right)^2\right] \quad (3.8)$$

For $\left(\dfrac{W}{H}\right) \geq 1$

$$\varepsilon_e = \frac{\varepsilon_r + 1}{2} + \frac{\varepsilon_r - 1}{2}\left[\left(1 + 12\left(\frac{H}{W}\right)\right)^{-0.5}\right] \quad (3.9)$$

This calculation ignores strip thickness and frequency dispersion, but their effects are negligible.

3.1.5 LOSSES IN MICROSTRIP ANTENNAS

Losses in microstrip line are due to conductor loss, radiation loss, and dielectric loss.

3.1.5.1 Conductor Loss

Conductor loss may be calculated by using Equation 3.10.

$$\alpha_c = 8.686 \log(R_S/(2WZ_0)) \quad \text{dB/length}$$

$$R_S = \sqrt{\pi f \mu \rho} \quad \text{Skin resistance} \quad (3.10)$$

Conductor losses may also be calculated by defining an equivalent loss tangent δ_c, given by $\delta_c = \delta_s/h$, where $\delta_s = \sqrt{\dfrac{2}{\omega \mu \sigma}}$. Where σ is the strip conductivity, h is the substrate height and μ is the free space permeability.

3.1.5.2 Dielectric Loss

Dielectric loss may be calculated by using Equation 3.11.

$$\alpha_d = 27.3 \frac{\varepsilon_r}{\sqrt{\varepsilon_{\text{eff}}}} \frac{\varepsilon_{\text{eff}} - 1}{\varepsilon_r - 1} \frac{tg\delta}{\lambda_0} \quad \text{dB/cm} \quad (3.11)$$

$$tg\delta = \text{dielectric loss coefficent}$$

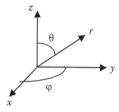

FIGURE 3.6   Coordinate system.

3.1.6 PATCH RADIATION PATTERN

The patch width, W, controls the antenna radiation pattern. The coordinate system is shown in Figure 3.6. The normalized radiation pattern is approximately given by

$$E_\theta = \frac{\sin\left(\dfrac{k_0 W}{2}\sin\theta\sin\varphi\right)}{\dfrac{k_0 W}{2}\sin\theta\sin\varphi}\cos\left(\dfrac{k_0 L}{2}\sin\theta\cos\varphi\right)\cos\varphi$$

(3.12)

$$k_0 = \frac{2\pi}{\lambda}$$

$$E_\varphi = \frac{\sin\left(\dfrac{k_0 W}{2}\sin\theta\sin\varphi\right)}{\dfrac{k_0 W}{2}\sin\theta\sin\varphi}\cos\left(\dfrac{k_0 L}{2}\sin\theta\cos\varphi\right)\cos\theta\sin\varphi$$

(3.13)

$$k_0 = \frac{2\pi}{\lambda}$$

The magnitude of the fields is given by

$$f(\theta,\varphi) = \sqrt{E_\theta^2 + E_\varphi^2}$$

(3.14)

3.2 TWO-LAYER STACKED MICROSTRIP ANTENNAS

Two-layer microstrip antennas was presented first in Refs. [1,5–8]. The major disadvantage of single-layer microstrip antennas is narrow bandwidth. By designing a double-layer microstrip antenna we can get a wider bandwidth.

In the first layer the antenna feed network and a resonator are printed. In the second layer the radiating element is printed. The electromagnetic field is coupled from the resonator to the radiating element. The resonator and the radiating element shape may be a square, rectangle, triangle, circle, or any arbitrary shape. The distance between the layers is optimized to obtain maximum bandwidth with the best antenna efficiency. The spacing between the layers may be air or a foam with low dielectric losses.

A circular polarization double-layer antenna was designed at 2.2 GHz. The resonator and the feed network were printed on a substrate with a relative dielectric constant of 2.5 with thickness of 1.6 mm. The resonator is a square microstrip resonator with dimensions $W = L = 45$ mm. The radiating element was printed on a substrate with a relative dielectric constant of 2.2 with thickness of 1.6 mm. The radiating element is a square patch with dimensions $W = L = 48$ mm. The patch was designed as a circular polarized antenna by connecting a 3 dB 90° branch coupler to antenna feed lines, as shown in Figure 3.7. The antenna bandwidth is 10% for voltage standing wave ratio (VSWR) better than 2:1. The measured antenna beam width is around 72°. The measured antenna gain is 7.5 dBi. Measured results of stacked microstrip antennas are listed in Table 3.1.

Results in Table 3.1 indicate that bandwidth of two-layer microstrip antennas may be around 9%–15% for VSWR better than 2:1. In Figure 3.7 a stacked microstrip antenna is shown. The antenna feed network is printed on a 1.6 mm thick FR4 dielectric substrate with dielectric constant of 4. The radiator is printed on a 1.6 mm thick RT/Duroid® 5880 dielectric substrate with dielectric constant of 2.2. The antenna electrical parameters were calculated and optimized by using ADS software. The dimensions of the microstrip stacked patch antenna shown in Figure 3.8 are $33 \times 20 \times 3.2$ mm. The computed S_{11} parameters are presented in Figure 3.9. The radiation pattern of the microstrip stacked patch is shown in Figure 3.10. The antenna bandwidth is around 5% for VSWR better than 2.5:1. The antenna bandwidth is improved to

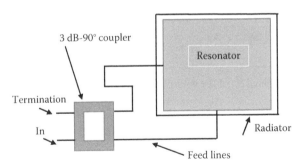

FIGURE 3.7   Circular polarized microstrip stacked patch antenna.

TABLE 3.1
Measured Results of Stacked Microstrip Antennas

Antenna	F (GHz)	Bandwidth (%)	Beam Width (°)	Gain (dBi)	Side Lobe (dB)	Polarization
Square	2.2	10	72	7.5	−22	Circular
Circular	2.2	15	72	7.9	−22	Linear
Annular disc	2.2	11.5	78	6.6	−14	Linear
Rectangular	2.0	9	72	7.4	−25	Linear
Circular	2.4	9	72	7	−22	Linear
Circular	2.4	10	72	7.5	−22	Circular
Circular	10	15	72	7.5	−25	Circular

FIGURE 3.8 A microstrip stacked patch antenna.

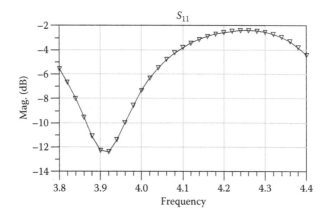

FIGURE 3.9 Computed S_{11} of the microstrip stacked patch.

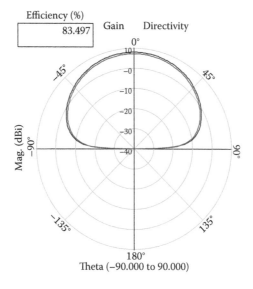

FIGURE 3.10 Radiation pattern of the microstrip stacked patch.

10% for VSWR better than 2.0:1 by adding 8 mm air spacing between the layers. The antenna beam width is around 72°. The antenna gain is around 7 dBi.

3.3   STACKED MONOPULSE Ku BAND PATCH ANTENNA

A monopulse double layer antenna was designed at 15 GHz. The monopulse double-layer antenna consists of four circular patch antennas as shown in Figure 3.11. The resonator and the feed network were printed on a 0.8 mm thick substrate with relative dielectric constant of 2.5. The resonator is a circular microstrip resonator with diameter $a = 4.2$ mm. The radiating element was printed on a 0.8 mm thick substrate with relative dielectric constant of 2.2. The radiating element is a circular microstrip patch with diameter $a = 4.5$ mm. The four circular patch antennas are connected to three 3 dB 180° rat-race couplers via the antenna feed lines, as shown in Figure 3.11. The comparator consists of three stripline 3 dB 180° rat-race couplers printed on a 0.8 mm thick substrate with relative dielectric constant of 2.2. The comparator has four output ports: a sum port Σ, difference port Δ, elevation difference port ΔE_1, and azimuth difference port ΔAz as shown in Figure 3.11. The antenna bandwidth is 10% for VSWR better than 2:1. The antenna beam width is around 36°. The measured antenna gain is around 10 dBi. The comparator losses are around 0.7 dB.

3.3.1   RAT-RACE COUPLER

A rat-race coupler is shown in Figure 3.12. The rat-race circumference is 1.5 wavelengths. The distance from A to Δ port is $3\lambda\backslash4$. The distance from A to Σ port is $\lambda\backslash4$. For an equal-split rat-race coupler, the impedance of the entire ring is fixed at $1.41 \times Z_0$,

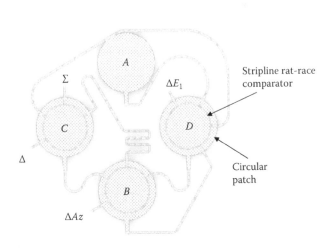

FIGURE 3.11   A microstrip stacked monopulse antenna.

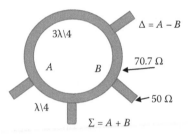

FIGURE 3.12 Rat-race coupler.

or 70.7 Ω for $Z_0 = 50$ Ω. For an input signal V, the outputs at ports 2 and 4 are equal in magnitude, but 180° out of phase.

3.4   LOOP ANTENNAS

Loop antennas are used as receive antennas in communication and medical systems. Loop antennas may be printed on a dielectric substrate or manufactured as a wired antenna. In this section several loop antennas are presented.

3.4.1   SMALL LOOP ANTENNA

A small loop antenna is shown in Figure 3.13. The shape of the loop antenna may be circular or rectangular. These antennas have low radiation resistance and high reactance. It is difficult to match the antenna to a transmitter. Loop antennas are most often used as receive antennas, where impedance mismatch loss can be accepted. Small loop antennas are used as field strength probes, in pagers, and in wireless measurements.

The loop lies in the x–y plane. The radius a of the small loop antenna is smaller than a wavelength ($a \ll \lambda$). A loop antenna electric field is given in Equation 3.17. A loop antenna magnetic field is given in Equations 3.15 and 3.16.

$$E_\phi = \eta \frac{(ak)^2 I_0 \sin\theta}{4r} e^{j(\omega t - \beta r)} \tag{3.15}$$

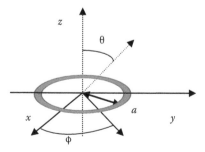

FIGURE 3.13   Small loop antenna.

$$H_\theta = -\frac{(ak)^2 I_0 \sin\theta}{4r} e^{j(\omega t - \beta r)} \tag{3.16}$$

The variation of the radiation pattern with direction is $\sin\theta$, the same as for a dipole antenna. The fields of a small loop have the E and H fields switched relative to that of a short dipole. The E field is horizontally polarized in the x–y plane.

The small loop is often referred to as the dual of the dipole antenna, because if a small dipole had magnetic current flowing (as opposed to electric current as in a regular dipole), the fields would resemble that of a small loop. The short dipole has a capacitive impedance (the imaginary part of the impedance is negative). The imped-ance of a small loop is inductive (positive imaginary part). The radiation resistance (and ohmic loss resistance) can be increased by adding more turns to the loop. If there are N turns of a small loop antenna, each with a surface area S, the radiation resistance for small loops can be approximated as given in Equation 3.17.

$$R_{rad} = \frac{31,329 N^2 S^2}{\lambda^4} \tag{3.17}$$

For a small loop, the reactive component of the impedance can be determined by finding the inductance of the loop. For a circular loop with radius a and wire radius r, the reactive component of the impedance is given by Equation 3.18.

$$X = 2\pi a f \mu \left(\ln\left(\frac{8a}{r}\right) - 1.75 \right) \tag{3.18}$$

Loop antennas behave better in the vicinity of the human body than dipole anten-nas. The reason is that the electric near fields in dipole antennas are very strong. For $r \ll 1$, the dominant component of the field varies as $1/r^3$. These fields are the dipole near fields. In this case the waves are standing waves and the energy oscillates in the antenna near zone and is not radiated to the open space. The real part of the Poynting vector is equal to zero. Near the body the electric fields decay rapidly. However, the magnetic fields are not affected near the body. The magnetic fields are strong in the near field of the loop antenna. These magnetic fields give rise to the loop antenna radiation. The loop antenna radiation near the human body is stronger than the dipole radiation near the human body. Several loop antennas are used as "wearable antennas."

3.4.2 Printed Loop Antenna

The diameter of a printed loop antenna is around a half wavelength. A loop antenna is dual to a half-wavelength dipole. Several loop antennas were designed for medi-cal systems at frequency ranges between 400 MHz and 500 MHz. In Figure 3.14 a printed loop antenna is presented. The antenna was printed on FR4 with 0.5 mm thickness. The loop diameter is 45 mm. The loop antenna VSWR is around 4:1. The printed loop antenna radiation pattern at 435 MHz is shown Figure 3.15. The loop antenna gain is around 1.8 dBi. An antenna with a tuning capacitor is shown in Figure 3.16. The loop antenna VSWR without the tuning capacitor was 4:1. This

FIGURE 3.14   Printed loop antenna.

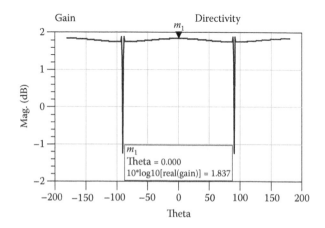

FIGURE 3.15   Printed loop antenna radiation pattern at 435 MHz.

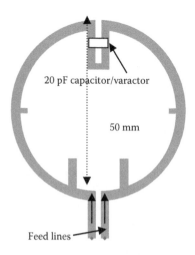

FIGURE 3.16   Tunable loop antenna without ground plane.

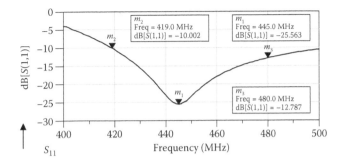

FIGURE 3.17 Computed S_{11} of loop antenna, without ground plane, with a tuning capacitor.

loop antenna may be tuned by adding a capacitor or varactor as shown in Figure 3.17. Matching stubs are employed to tune the antenna to the resonant frequency. Tuning the antenna allows us to work in a wider bandwidth as shown in Figure 3.18. Loop antennas are used as receive antennas in medical systems. The loop antenna radiation pattern on the human body is shown Figure 3.19. The computed 3D radiation pattern and the coordinate used in this chapter are shown in Figure 3.19.

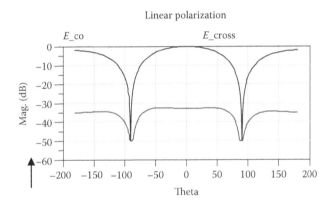

FIGURE 3.18 Radiation pattern of loop antenna on the human body.

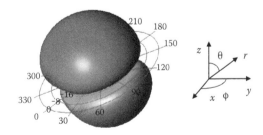

FIGURE 3.19 Loop antenna 3D radiation pattern.

3.4.3   Radio Frequency Identification Loop Antennas

Radio frequency identification (RFID) loop antennas are widely used. Several RFID loop antennas are presented in Ref. [10]. RFID loop antennas have low efficiency and narrow bandwidth. As an example, the measured impedance of a square four-turn loop at 13.5 MHz is $0.47 + j107.5\ \Omega$. A matching network is used to match the antenna to 50 Ω. The matching network consists of a 56 pF shunt capacitor, 1 kΩ shunt resistor, and another 56 pF capacitor. This matching network has narrow band-width. The antenna is printed on a FR4 substrate. The antenna dimensions are 32 × 52.4 × 0.25 mm. The antenna layout is shown in Figure 3.20. A photo of the RFID antenna is shown in Figure 3.21. S_{11} results of the printed loop antenna are shown in Figure 3.22. The antenna S_{11} parameter is better than –9.5 dB without an external matching network. The computed radiation pattern is shown in Figure 3.23. The RFID antenna beam width is around 160°.

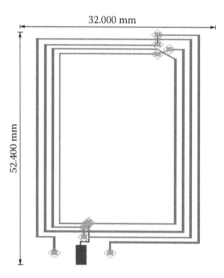

FIGURE 3.20   A square four turn loop antenna.

FIGURE 3.21   Four turn 13.5-MHz loop rectangular antenna.

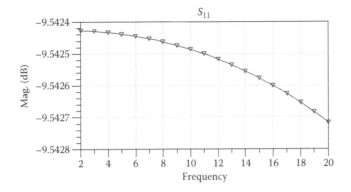

FIGURE 3.22   RFID loop antenna computed S_{11} results.

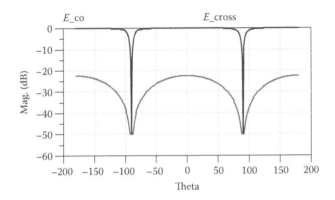

FIGURE 3.23   RFID loop antenna radiation pattern.

3.4.4   NEW LOOP ANTENNA WITH GROUND PLANE

A new loop antenna with ground plane has been designed on Kapton substrates with a relative dielectric constant of 3.5 and thickness of 0.25 mm. The antenna is shown in Figure 3.24. Matching stubs are employed to tune the antenna to the resonant frequency. The loop antenna with ground plane diameter is 45 mm. The antenna was designed by employing ADS software. The antenna center frequency is 427 MHz. The antenna bandwidth for VSWR better than 2:1 is around 12%, as shown in Figure 3.25. The printed loop antenna radiation pattern at 435 MHz is shown in Figure 3.26. The loop antenna gain is around 1.8 dBi. The loop antenna with ground plane beam width is around 100°.

A printed loop antenna with ground plane with shorter tuning stubs is shown in Figure 3.27. The loop antenna with ground with shorter tuning stubs plane diameter is 45 mm. The antenna center frequency is 438 MHz. The antenna bandwidth for VSWR better than 2:1 is around 12%, as shown in Figure 3.28. A printed loop antenna with shorter tuning stubs radiation pattern at 435 MHz is shown Figure 3.29.

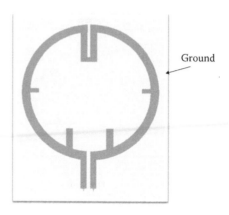

FIGURE 3.24 Printed loop antenna with ground plane.

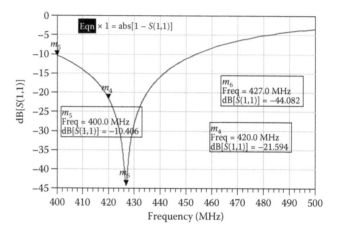

FIGURE 3.25 Computed S_{11} of loop antenna with ground plane.

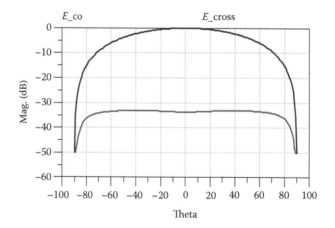

FIGURE 3.26 Radiation pattern of loop antenna, Figure 3.24, with ground plane.

FIGURE 3.27   Printed loop antenna with ground plane and short tuning stubs.

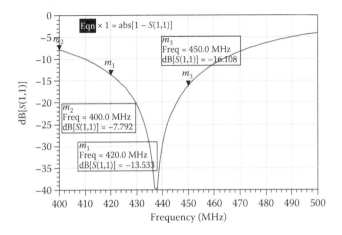

FIGURE 3.28   Radiation pattern of loop antenna with ground plane.

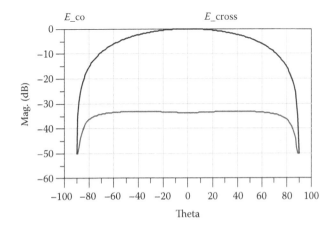

FIGURE 3.29   Radiation pattern of loop antenna, Figure 3.27, with ground plane.

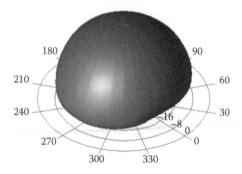

FIGURE 3.30   3D radiation pattern of loop antenna with ground plane.

The loop antenna gain is around 1.8 dBi. The loop antenna with ground plane beam width is around 100°. A printed loop antenna with a shorter tuning stub 3D radiation pattern at 430 MHz is shown Figure 3.30.

3.5   WIRED LOOP ANTENNA

A wire seven-turn loop antenna is shown in Figure 3.31. The antenna length $l = 4.5$ mm. The loop diameter is 3.5 mm. Equation 3.19 is an approximation to calculate the inductance value for an air coil loop antenna with N turns, diameter r, and length l.

$$L(nH) = \frac{3.94 r_{mm} N^2}{0.9\dfrac{\ell}{r}+1} = \frac{0.1 r_{mil} N^2}{0.9\dfrac{\ell}{r}+1} \tag{3.19}$$

An approximation to calculate the inductance value for an air coil loop antenna with N turns, diameter r, and length l is given in Equation 3.20.

$$L(nH) \approx \frac{r_{mm}^2 N^2}{2l + r} \tag{3.20}$$

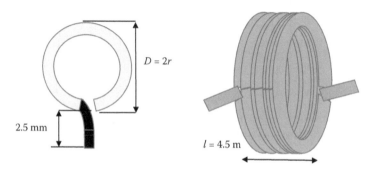

FIGURE 3.31   Wire seven-turn loop antenna.

FIGURE 3.32 Wire loop antenna on PCB board.

FIGURE 3.33 Seven-turn loop antenna on PCB board.

For length l = 4.5 mm, wire diameter 0.6 mm, loop diameter 3.5 mm, and N = 7 the inductance is around L = 52 nH. The quality factor of this seven-turn wire loop is around 100. For length l = 1.7 mm, wire diameter 0.6 mm, loop diameter 6.5 mm, and N = 2 the inductance is around L = 45 nH. For length l = 2 mm, wire diameter 0.6 mm, loop diameter 7.62 mm, and N = 2 the inductance is around L = 42.1 nH. For length l = 0.5 mm, wire diameter 0.6 mm, loop diameter 7 mm, and N = 2 the inductance is around L = 87 nH. For length l = 2.5 mm, wire diameter 0.6, loop diameter 5 mm, and N = 2 the inductance is around L = 20.7 nH.

The wire loop antenna has very low radiation efficiency. The amount of power radiated is a small fraction of the input power. The antenna efficiency is around 0.01%, −41 dB. The ratio between the antenna's dimension and wavelength is around 1:100. Small antennas are characterized by radiation resistance R_r. The radiated power is given by I^2R_r, where I is the current through the coil. Remember that the current through the coil is Q times the current through the antenna. For example, for current of 2 mA and Q of 20, if R_r = 10^{-3} Ω, the radiated power is around −32 dBm.

Figure 3.32 presents a wire loop antenna with 2.5 turns on a PCB board. Figure 3.33 presents a wire loop antenna with seven turns on a PCB board.

3.6 RADIATION PATTERN OF A LOOP ANTENNA NEAR A METAL SHEET

An E- and H-plane 3D radiation pattern of a wire loop antenna in free space is shown in Figure 3.34. An E- and H-plane 3D radiation pattern of a wire loop antenna near

FIGURE 3.34 *E*- and *H*-plane radiation pattern of loop antenna in free space.

a metal sheet as shown in Figure 3.35 was computed by employing HFSS software. Figure 3.36 presents the *E*- and *H*-plane radiation pattern of loop antenna for distance of 30 cm from a metal sheet. We can see that a metal sheet in the vicinity of the antenna split the main beam and creates holes of around 20 dB in the radiation pattern. Figure 3.37 presents the *E*- and *H*-plane radiation pattern of loop antenna located 10 cm from a metal sheet as presented in Figure 3.38. We can see that a metal sheet in the vicinity of the antenna split the main beam and creates holes up to 15 dB in the radiation pattern.

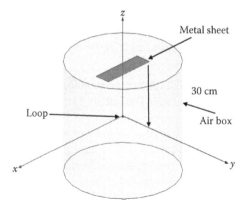

FIGURE 3.35 Loop antenna located near a metal sheet.

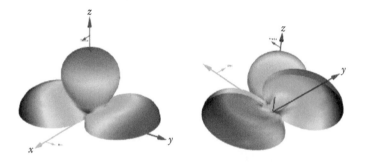

FIGURE 3.36 *E*- and *H*-plane radiation pattern of a loop antenna for a distance of 30 cm from a metal sheet.

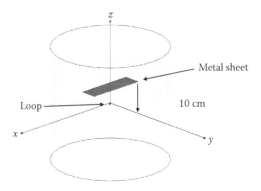

FIGURE 3.37 Loop antenna located 10 cm from a metal sheet.

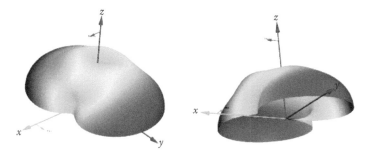

FIGURE 3.38 *E*- and *H*-plane radiation pattern of a loop antenna for distance of 10 cm from a metal sheet.

3.7 PLANAR INVERTED-F ANTENNA

The planar inverted-F antenna (PIFA) possesses attractive features such as low profile, small size, and low fabrication costs [16–18]. The PIFA antenna bandwidth is higher than the bandwidth of the conventional patch antenna, because the PIFA antenna thickness is higher than the thickness of patch antennas. The conventional PIFA antenna is a grounded quarter-wavelength patch antenna. The antenna consists of ground-plane a top plate radiating element printed on a dielectric substrate, feed wire, and a shorting plate or short circuited via holes from the radiating element to the ground-plane as shown in Figure 3.39.

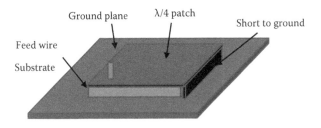

FIGURE 3.39 Conventional PIFA antenna.

The patch is shorted at the end. The fringing fields, which are responsible for radiation, are shorted on the far end, so only the fields nearest the transmission line radiate. Consequently, the gain is reduced, but the patch antenna maintains the same basic properties as a half-wavelength patch. However, the antenna length is reduced by 50%. The feed location may be placed between the open and the shorted end. The feed location controls the antenna input impedance.

3.7.1 GROUNDED QUARTER-WAVELENGTH PATCH ANTENNA

A grounded quarter-wavelength patch antenna was designed on a 1.6 mm thick FR4 substrate with relative dielectric constant of 4.5 at 3.85 GHz. The antenna is shown in Figure 3.40. The antenna was designed by using ADS software. The antenna dimensions are 34 × 17 × 1.6 mm. S_{11} results of the antenna are shown in Figure 3.41. The antenna bandwidth is around 6% for VSWR better than 3:1 without a matching network. The radiation pattern is shown in Figure 3.42. The antenna beam width is around 76°. The grounded quarter-wavelength patch antenna gain is around 6.7 dBi. The antenna efficiency is around 92%.

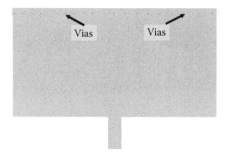

FIGURE 3.40 Grounded quarter-wavelength patch antenna.

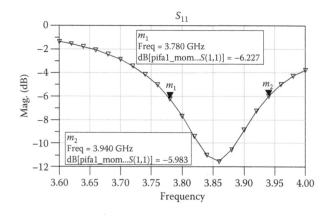

FIGURE 3.41 Grounded quarter-wavelength patch antenna S_{11} results.

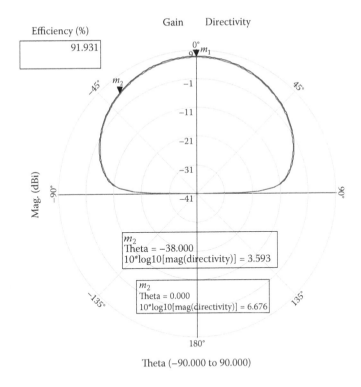

FIGURE 3.42   Grounded quarter-wavelength patch antenna radiation pattern.

3.7.2   A NEW DOUBLE-LAYER PIFA ANTENNA

A new double-layer PIFA antenna was designed. The first layer is a grounded quarter-wavelength patch antenna printed on a 1.6 mm thick FR4 substrate with relative dielectric constant of 4.5. The second layer is a rectangular patch antenna printed on a 1.6 mm thick Duroid substrate with relative dielectric constant of 2.2. The antenna is shown in Figure 3.43. The antenna was designed by employing ADS software. The

FIGURE 3.43   Double-layer PIFA antenna.

antenna dimensions are 34 × 17 × 3.2 mm. S_{11} results of the antenna are shown in Figure 3.44. The antenna is a dual-band antenna. The first resonant frequency is 3.48 GHz. The second resonant frequency is 4.02 GHz. The radiation pattern is shown in Figure 3.45. The antenna beam width is around 74°. The antenna gain is around 7.4 dBi. The antenna efficiency is around 83.4%.

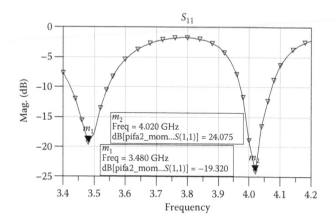

FIGURE 3.44 Double-layer PIFA antenna S_{11} results.

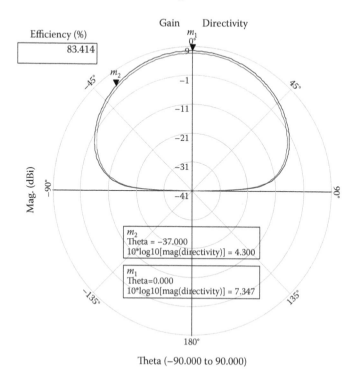

FIGURE 3.45 Double-layer PIFA antenna radiation pattern.

REFERENCES

1. James, J. R., Hall, P. S., and Wood, C. *Microstrip Antenna Theory and Design*. The Institution of Engineering and Technology, London, 1981.
2. Sabban, A., and Gupta, K. C. Characterization of Radiation Loss from Microstrip Discontinuities Using a Multiport Network Modeling Approach. *IEEE Transactions on Microwave Theory and Techniques*, 39(4):705–712, 1991.
3. Sabban, A. *Multiport Network Model for Evaluating Radiation Loss and Coupling Among Discontinuities in Microstrip Circuits*. PhD Thesis, University of Colorado at Boulder, January 1991.
4. Kathei, P. B., and Alexopoulos, N. G. Frequency-Dependent Characteristic of Microstrip, Discontinuities in Millimeter-Wave Integrated Circuits. *IEEE Transactions on Microwave Theory and Techniques*, MTT-33, 1029–1035, 1985.
5. Sabban, A. A New Wideband Stacked Microstrip Antenna. In *IEEE Antenna and Propagation Symposium*, Houston, TX, June 1983.
6. Sabban, A. Wideband Microstrip Antenna Arrays. In *IEEE Antenna and Propagation Symposium MELCOM*, Tel Aviv, Israel, June 1981.
7. Sabban, A. Microstrip Antennas. In *IEEE Symposium*, Tel Aviv, Israel, October 1979.
8. Sabban, A., and Navon, E. A mm-Waves Microstrip Antenna Array. In *IEEE Symposium*, Tel Aviv, Israel, March 1983.
9. de Lange, G. et al. A 3*3 mm-Wave Micro Machined Imaging Array with Sis Mixers. *Applied Physics Letters*, 75(6):868–870, 1999.
10. Rahman, A. et al. Micromachined Room Temperature Microbolometers for mm-Wave Detection. *Applied Physics Letters*, 68 (14):2020–2022, 1996.
11. Milkov, M. M. *Millimeter-Wave Imaging System Based on Antenna-Coupled Bolometer*. MSc. Thesis, UCLA, 2000.
12. Jack, M. D. et al. US Patent 6329655, 2001.
13. Sinclair, G. N. et al. Passive Millimeter Wave Imaging in Security Scanning. *Proceedings of SPIE*, 4032:40–45, 2000.
14. Kompa, G., and Mehran, R. Planar Waveguide Model for Computing Microstrip Components. *Electronic Letters*, 11(9):459–460, 1975.
15. Luukanen, A. et al. US Patent 6242740, 2001.
16. Wang, F., Du, Z., Wang, Q., and Gong, K. Enhanced-Bandwidth PIFA with T-shaped Ground Plane. *Electronic Letters*, 40(23):1504–1505, 2004.
17. Kim, B., Hoon, J., and Choi, H. Small Wideband PIFA for Mobile Phones at 1800 MHz. In *Vehicular Technology Conference*, Milan, Italy, Vol. 1, pp. 27–29, May 2004.
18. Kim, B., Park, J., and Choi, H. Tapered Type PIFA Design for Mobile Phones at 1800 MHz. In *Vehicular Technology Conference*, Stockholm, Sweden, Vol. 2, pp. 1012–1014, April 2005.

4 Antenna Array

4.1 INTRODUCTION

An array of antenna elements is a set of antennas used for transmitting or receiving electromagnetic waves. An array of antennas is a collection of N similar radiators with the same three-dimensional radiation pattern. All the radiating elements are fed with the same frequency and with a specified amplitude and phase relationship for the drive voltage of each element. The array functions as a single antenna, generally with higher antenna gain than would be obtained from the individual elements. In antenna arrays electromagnetic wave interference is used to enhance the electromagnetic signal in one desired direction at the expense of other directions. It may also be used to null the radiation pattern in one particular direction. Antenna arrays theory is presented in Ref. [1]. Several printed arrays and gain limitations of printed arrays are presented in Refs. [2–6]. Analysis and computations of losses in microstrip lines are given in Refs. [2–6]. Millimeter (MM) wave arrays are presented in Refs. [6–9].

4.2 ARRAY RADIATION PATTERN

The polar radiation pattern of a single element is called the element pattern (EP). The array pattern is the polar radiation pattern that would result if the elements were replaced by isotropic radiators, with the same amplitude and phase of excitation as the actual elements, and spaced at points on a grid corresponding to the far field phase centers of the radiators. If we assume that all the polar radiation patterns of the elements taken individually are identical (within a certain tolerance) and that the patterns are all aligned in the same direction in azimuth and elevation, then the total array antenna pattern is obtained by multiplying the array factor, AF, by the element pattern. The total array antenna pattern, ET, is $ET = AF \cdot EP$.

The radiated field strength at a certain point in space in the far field is calculated by adding the contributions of each element to the total radiated fields. The summation of the contribution of each element to the total radiated fields is called the array factor, AF. The field strengths fall off as $1/r$, where r is the distance from the antenna to the field point as shown in Figure 4.1. We must take into account the amplitude and phase angle of the radiator excitation, and also the phase delay that is due to the time it takes the signal to get from the source to the field point. This phase delay is expressed as $2\pi r/\lambda$, where λ is the free space wavelength of the electromagnetic wave.

Contours of equal field strength may be interpreted as an amplitude polar radiation pattern. Contours of the squared modulus of the field strength may be interpreted as a power polar radiation pattern. Figure 4.1 presents a four-element array. The array

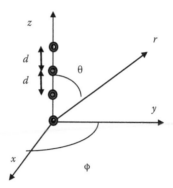

FIGURE 4.1   Coordinate system for external field calculations.

factor of N element array is given in Equation 4.1. β represents the phase difference between the elements in the array. Figure 4.2 presents a two-element array.

$$AF = 1 + e^{j\varphi} + e^{j2\varphi} + \ldots + e^{j\varphi(N-1)} = \sum_{n=1}^{N} e^{j\varphi(n-1)} \tag{4.1}$$

where $\varphi = kd \cos\theta + \beta$, with $k = 2\pi/\lambda$. The series summation is given in Equation 4.2.

$$AF = \frac{\sin\left(\dfrac{N\varphi}{2}\right)}{\sin\left(\dfrac{\varphi}{2}\right)} \tag{4.2}$$

The array factor will be zero when $\sin\left(\dfrac{N\varphi}{2}\right) = 0$. The array nulls will occur as given in Equation 4.3.

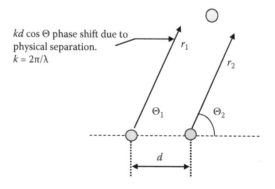

FIGURE 4.2   Two-element array.

$$\theta_n = \cos^{-1}\left[\frac{\lambda}{2\pi d}\left(-\beta \pm \frac{2n}{N}\pi\right)\right] \quad n \neq N = 1,2,3\ldots \tag{4.3}$$

For $\varphi = \pm 2m\pi$ the array maximum level is given in Equation 4.4.

$$\theta_m = \cos^{-1}\left[\frac{\lambda}{2\pi d}(-\beta \pm 2m\pi)\right] \quad m = 0,1,2,3\ldots \tag{4.4}$$

The array 3 dB beam width is given in Equation 4.5.

$$\theta_{3\,\text{dB}} = \cos^{-1}\left[\frac{\lambda}{2\pi d}\left(-\beta \pm \frac{2.782}{N}\right)\right] \tag{4.5}$$

The peak value of the side lobe is given in Equation 4.6.

$$\theta_{\text{SL}} = \cos^{-1}\left[\frac{\lambda}{2\pi d}\left(-\beta \pm \frac{3\pi}{N}\right)\right] \tag{4.6}$$

The side-lobe level for $\beta = 0$ is -13.46 dB, as calculated in Equation 4.7.

$$AF_{\text{SL}} = 20\log_{10}\left(\frac{2}{3\pi}\right) = -13.46 \text{ dB} \tag{4.7}$$

4.3 BROADSIDE ARRAY

In a broadside array the main beam is perpendicular to the array. The array will radiate maximum energy perpendicular to the array if all elements in the array are fed with the same amplitude and phase level, $\beta = 0$. The array factor will be zero when $\sin\left(\frac{N\varphi}{2}\right)$. The array nulls will occur as given in Equation 4.8.

$$\theta_n = \cos^{-1}\left(\pm\frac{n\lambda}{Nd}\right) \quad n \neq N = 1,2,3\ldots \tag{4.8}$$

For $\varphi = \pm 2m\pi$ the array maximum level is given in Equation 4.9.

$$\theta_m = \cos^{-1}\left(\frac{m\lambda}{d}\right) \quad m = 0,1,2,3\ldots \tag{4.9}$$

The array 3 dB beam width is given in Equation 4.10 when $\pi d/\lambda \ll 1$.

$$\theta_{3\,\text{dB}} = \cos^{-1}\left(\frac{1.391\lambda}{\pi Nd}\right) \quad \pi d/\lambda \ll 1 \tag{4.10}$$

The peak value of the side lobe is given in Equation 4.11.

$$\theta_{SL} = \cos^{-1}\left[\frac{\lambda}{2d}\left(\pm\frac{(2s+1)}{N}\right)\right] \quad s = 0,1,2,3\ldots \tag{4.11}$$

4.4   END-FIRE ARRAY

In end-fire array the main beam is in the direction of the array axis. The array will radiate maximum energy in the direction of the array axis. The array will radiate maximum energy in the direction $\theta = 0$ if $\beta = -kd$. The array will radiate maximum energy in the direction $\theta = 180$ if $\beta = kd$.

The array nulls will occur as given in Equation 4.12.

$$\theta_n = \cos^{-1}\left(1-\frac{n\lambda}{Nd}\right) \quad n \neq N = 1,2,3\ldots \tag{4.12}$$

For $\varphi = \pm 2m\pi$ the array maximum level is given in Equation 4.13.

$$\theta_m = \cos^{-1}\left(1-\frac{m\lambda}{d}\right) \quad m = 0,1,2,3\ldots \tag{4.13}$$

The array 3 dB beam width is given in Equation 4.14 when $\pi d/\lambda \ll 1$.

$$\theta_{3\,dB} = \cos^{-1}\left(1-\frac{1.391\lambda}{\pi Nd}\right) \quad \pi d/\lambda \ll 1 \tag{4.14}$$

The peak value of the side lobe is given in Equation 4.15.

$$\theta_{SL} = \cos^{-1}\left[\frac{\lambda}{2d}\left(1-\frac{(2s+1)}{N}\right)\right] \quad s = 0,1,2,3\ldots \tag{4.15}$$

4.5   PRINTED ARRAYS

A parallel feed network of microstrip antenna array is shown in Figure 4.3a. A parallel series feed network of microstrip antenna array is shown in Figure 4.3b.

Array directivity, D, may be written as $D = D_0 = U_{max}/U_0 \sim AF_{max}^2 = N$.

Half-power beam width may be written as HPBW $= 2 \times (90 - \arccos(1.39\lambda/\pi Nd))$. In Table 4.1 array directivity and beam width as a function of number of element are listed. The radiation pattern of a 16 broadside element array is shown in Figure 4.4. The array directivity is 12 dB.

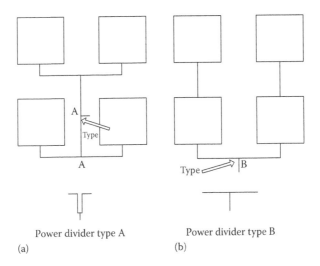

FIGURE 4.3 Configuration of microstrip antenna array. (a) Parallel feed network. (b) Parallel series feed network.

TABLE 4.1
Array Directivity as a Function of Number of Element

N Elements	D_0 (dB)	HPBW ($\theta_{3\,dB}$)
4	6	26°
8	9	10°
12	10.8	9°
16	12	7°
32	15	3.5°
64	18.1	1.75°

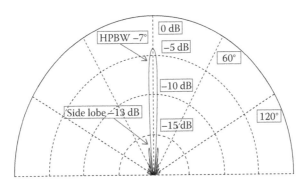

FIGURE 4.4 Radiation pattern of a broadside 16-element array.

4.6 STACKED MICROSTRIP ANTENNA ARRAYS

A stacked Ku band 16-element microstrip antenna array was designed at 14.5 GHz, as shown in Figure 4.5. The resonator and the feed network were printed on a substrate with relative dielectric constant of 2.5 with thickness of 0.5 mm. The resonator is a circular microstrip resonator with diameter $a = 4.2$ mm. The radiating element was printed on a substrate with relative dielectric constant of 2.2 with thickness of 0.5 mm. The distance between the radiating elements is around 0.75 λ. The array dimensions are $125 \times 40 \times 1$ mm. The antenna bandwidth is 10% for voltage standing wave ratio (VSWR) better than 2:1. The antenna beam width is around 10×25. The measured antenna gain is around 18.5 dBi. Losses in the feed network are around 1 dB.

A stacked Ku band 32-element microstrip antenna array was designed at 14.5 GHz as shown in Figure 4.6. The resonator and the feed network were printed on a 0.5 mm thick substrate with relative dielectric constant of 2.5. The resonator is a circular microstrip resonator with diameter $a = 4.2$ mm. The radiating element was printed on a 0.5 mm thick substrate with relative dielectric constant of 2.2. The distance between the radiating elements is around 0.75 λ. The array dimensions are $125 \times 125 \times 1$ mm. The antenna bandwidth is 10% for VSWR better than 2:1. The antenna beam width is around 10×20. The measured antenna gain is around 20.5 dBi. Losses in feed network are around 1.5 dB. Measured results of several stacked microstrip antenna arrays are listed in Table 4.2. The measured gain of the 64-element array at 34 GHz is around 23.5 dBi.

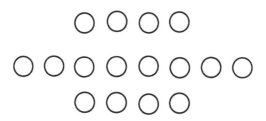

FIGURE 4.5   Stacked Ku band 16-element microstrip antenna array.

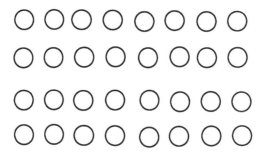

FIGURE 4.6   Stacked Ku band 32-element microstrip antenna array.

TABLE 4.2
Measured Results of Stacked Microstrip Antenna Array

Array	F (GHz)	Bandwidth (%)	Beam Width (°)	Gain dBi	Dimensions (mm)
16	14.5	10	10×25	18.5	$125 \times 40 \times 1$
32	14.5	10	10×20	20.5	$125 \times 125 \times 1$
16	10	13.5	23×23	17.0	$60 \times 60 \times 1.6$
64	34	10	10×10	23.5	$55 \times 55 \times 0.5$

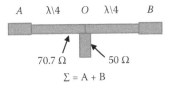

FIGURE 4.7 Power combiner/splitter.

A basic configuration of a power combiner/splitter is shown in Figure 4.7. This configuration of a power combiner/splitter is used in the feed network of several printed arrays. The power combiner/splitter consists of two sections of quarter-wavelength transformer. The distance from A to O is $\lambda\backslash4$. The impedance at point O is 100 Ω. For an equal-split splitter the impedance of the quarter wavelength trans-former is $1.41 \times Z_0$, or 70.7 Ω for $Z_0 = 50$ Ω. For an input signal V, the outputs at ports A and B are equal in magnitude and in phase.

4.7 Ka BAND MICROSTRIP ANTENNA ARRAYS

Microstrip antenna arrays with integral feed networks may be broadly divided into arrays fed by parallel feeds and series fed arrays. Usually series fed arrays are more efficient than parallel fed arrays. However, parallel fed arrays have a well-controlled aperture distribution. Two Ka band microstrip antenna arrays that consist of 64 radi-ating elements have been designed on a 10 mil Duroid® substrate with $\varepsilon_r = 2.2$. The first array uses a parallel feed network and the second uses a parallel series feed net-work as shown in Figure 4.8a and b. Comparison of the performance of the arrays is given in Table 4.3. Results given in Table 4.3 verify that the parallel series fed array is more efficient than the parallel fed array because of minimization of the number of discontinuities in the parallel series feed network.

The parallel series fed array has been modified by using a 5-centimeter coaxial line to replace the same length of microstrip line. Results given in Table 4.3 indicate that the efficiency of the parallel series fed array that incorporates a coaxial line in the feed network is around 67.6% because of minimization of the microstrip line length. Two microstrip antenna arrays that consist of 256 radiating elements have been designed. In the first array, type A, as shown in Figure 4.9a, using power divider

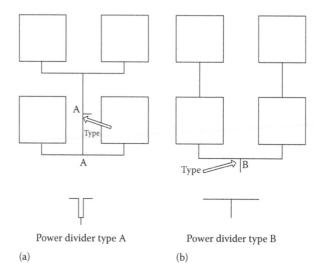

Power divider type A   Power divider type B

(a)   (b)

FIGURE 4.8 Confirmation of 64-element microstrip antenna array. (a) Parallel feed network. (b) Parallel series feed network.

TABLE 4.3
Performance of 64-Element Microstrip Antenna Arrays

Parameter	Corporate Feed	Parallel Feed	Corporate Feed Mix
Number of elements	64	64	64
Beam width (°)	8.5	8.5	8.5
Computed gain (dBi)	26.3	26.3	26.3
Microstrip line loss (dB)	1.1	1.2	0.5
Radiation losses (dB)	0.7	1.3	0.7
Mismatch loss (dB)	0.5	0.5	0.5
Expected gain	24.0 dBi	23.3 dBi	24.6 dBi
Efficiency (%)	58.9	50.7	67.6

type A minimizes the number of microstrip discontinuities. The second array, type B, as shown in Figure 4.9b, incorporates more bend discontinuities in the feeding network. The performances of the arrays are compared in Table 4.4. The type A array with 256 radiating elements has been modified by using a 10-centimeter coaxial line to replace the same length of microstrip line. The performance comparison of the arrays given in Table 4.2 shows that the gain of the modified array has increased by 1.6 dB. Results given in Table 4.4 verify that the type A array is more efficient than the type B array because of minimization of the number of bend discontinuities in the type A array feed network. The measured gain results are very close to the computed gain results and verify the loss computation.

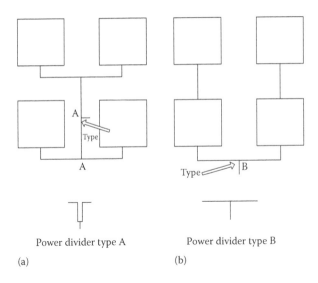

FIGURE 4.9 Configuration of 256-element microstrip antenna array. (a) Parallel feed network. (b) Parallel series feed network.

TABLE 4.4
Performance of 256-Element Microstrip Antenna Arrays

Parameter	Type A	Type B	Type C
Number of elements	256	256	256
Beam width (°)	4.2	4.2	4.2
Computed gain (dBi)	32	32	32
Microstrip line loss (dB)	3.1	3.1	1.5
Radiation losses (dB)	1	1.9	1
Mismatch loss (dB)	0.5	0.5	0.5
Expected gain (dBi)	27.43	26.5	29.03
Efficiency (%)	34.9	28.2	50.47

4.8   SERIES FED MICROSTRIP ARRAYS

In series fed microstrip arrays the patches are attached periodically to microstrip transmission line. The distance between the radiating elements d is around half a wavelength and is chosen to obtain the required phase distribution across the array aperture.

An S band series fed microstrip arrays four, eight, and sixteen elements were designed. Figures 4.10 and 4.11 presents series fed microstrip arrays with four and eight elements respectively. The resonator and the feed network were printed on a 1.6 mm

FIGURE 4.10   Four-element series fed microstrip array.

FIGURE 4.11   Eight-element series fed microstrip array.

thick substrate with relative dielectric constant of 2.25. The resonator is a rectangular patch resonator with dimension of 38.2 × 38.2 mm. The distance between the array elements, d, is 96 mm.

S_{11} results of the four-element series fed patch array are shown in Figure 4.12. The array resonant frequency is 2.465 GHz. The array impedance as a function of frequency is shown in Figure 4.13. The array impedance at 2.465 GHz is 46.2 Ω. The impedance of a single patch in the array is 46.2 × 4 = 184.8 Ω. The equivalent model of N elements series fed array is N impedances connected in parallel.

The radiation pattern of the four-element series fed array at 2.4 GHz is shown in Figure 4.14. The array beam width is around 18°. The antenna gain is around 14 dBi with 5° scanning angle. The first side-lobe level is around –13.6 dB. S_{11} results of the four-element series fed patch array with a distance of 63 mm between the array elements are shown in Figure 4.15. The array resonant frequency is 2.42 GHz. The radiation pattern of this four-element series fed array at 2.4 GHz is shown in Figure 4.16. S_{11} results of the eight-element series fed patch array with d = 96 mm are shown in Figure 4.17. The radiation pattern at 2.41 GHz of the eight-element

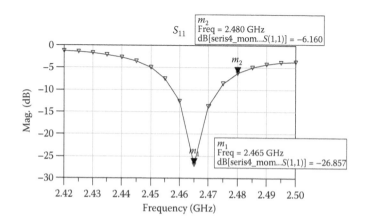

FIGURE 4.12 S_{11} results of the four-element series fed patch array.

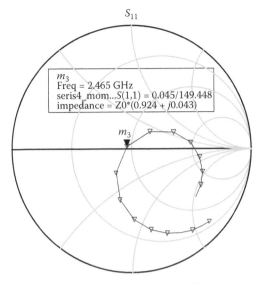

Freq (2.420 GHz to 2.500 GHz)

FIGURE 4.13 The four-element series fed array impedance as a function of frequency.

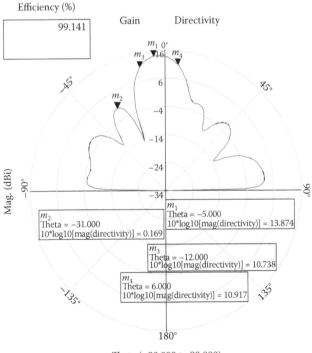

Theta (−90.000 to 90.000)

FIGURE 4.14 Radiation pattern of the four-element series fed array.

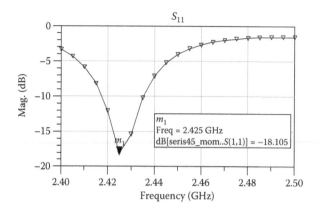

FIGURE 4.15 S_{11} results of the four-elements series fed patch array with $d = 63$ mm.

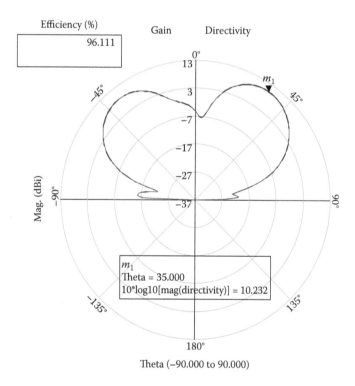

FIGURE 4.16 Radiation pattern of the four-element series fed array with $d = 63$ mm.

series fed array with $d = 96$ mm is shown in Figure 4.18. The array beam width is around 13°. The antenna gain is around 14.4 dBi with 1° scanning angle. The radiation pattern of the eight-element series fed array at 2.44 GHz is shown in Figure 4.19. The array beam width is around 12°. The antenna gain is around 17 dBi with 3° scanning angle. The first side-lobe level is around 13 dB. The radiation

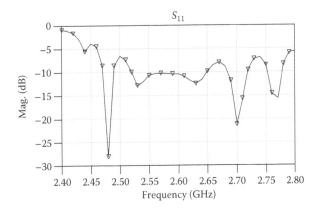

FIGURE 4.17 S_{11} results of the eight-element series fed patch array with $d = 96$ mm.

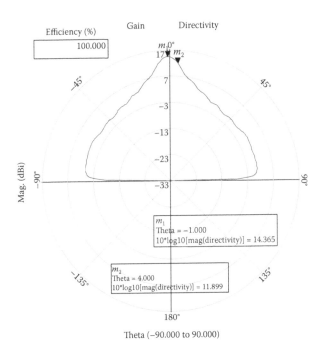

FIGURE 4.18 Radiation pattern of the eight-element series fed array with $d = 96$ mm.

pattern of the eight-element series fed array at 2.48 GHz is shown in Figure 4.20. The array beam width is around 10°. The antenna gain is around 16.5 dBi with 8° scanning angle. The first side-lobe level is around −5 dB. Data for scanning angle as a function of frequency of the eight-element series fed array are summarized in Table 4.5.

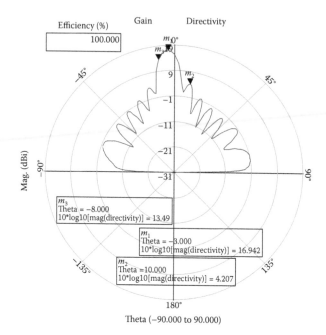

FIGURE 4.19   Radiation pattern of the eight-element series fed array at 2.44 GHz.

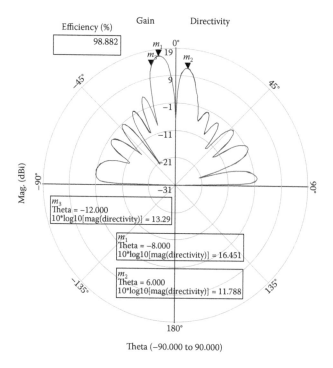

FIGURE 4.20   Radiation pattern of the eight-element series fed array at 2.48 GHz.

TABLE 4.5

**Scanning Angle as a Function of Frequency
of the Eight-Element Series Fed Array**

Frequency (GHz)	2.4	2.41	2.44	2.48	2.53
Scanning angle°	0	1	3	8	12

4.9   STACKED SERIES FED MICROSTRIP 8-ELEMENT ARRAY

An S band stacked series fed eight-element patch array was designed as shown in Figure 4.21. The resonator and the feed network were printed on a 1.6 mm thick substrate with relative dielectric constant of 2.25. The resonator is a rectangular patch resonator with dimension of 38.2 × 38.2 mm. The distance between the array elements, d, is 96 mm.

In the second layer the radiating element was printed on a 1.6 mm thick substrate with relative dielectric constant of 2.25. The electromagnetic field is coupled from the resonator to the radiating element. The radiating element is a square patch with dimensions $W = L = 40$ mm. S_{11} results of the stacked eight-element series fed patch array with $d = 96$ mm are shown in Figure 4.22. The antenna bandwidth for VSWR better than 3:1 is around 10%. The radiation pattern at 2.28 GHz of the stacked eight-element series fed array with $d = 96$ mm is shown in Figure 4.23. The array beam width is around 14°. The antenna gain is around 13 dBi with 2° scanning angle.

The radiation pattern at 2.32 GHz of the stacked eight-element series fed array with $d = 96$ mm is shown in Figure 4.24. The array beam width is around 10°. The antenna gain is around 15.6 dBi with 5° scanning angle. The radiation pattern at 2.4 GHz of the stacked eight elements series fed array with $d = 96$ mm is shown in Figure 4.25. The array beam width is around 13°. The antenna gain is around 13.6 dBi with 14° scanning angle.

FIGURE 4.21   Eight-element stacked series fed microstrip array.

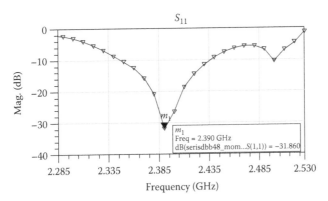

FIGURE 4.22 S_{11} results of the stacked eight-element series fed patch array.

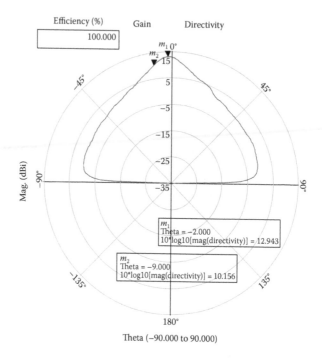

FIGURE 4.23 Radiation pattern of the stacked eight-element series fed array at 2.28 GHz.

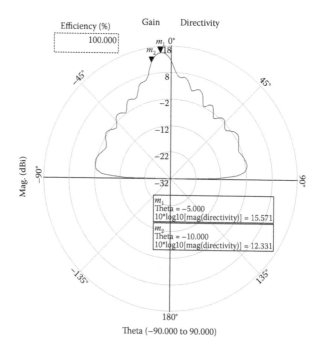

FIGURE 4.24 Radiation pattern of the stacked eight-element series fed array at 2.32 GHz.

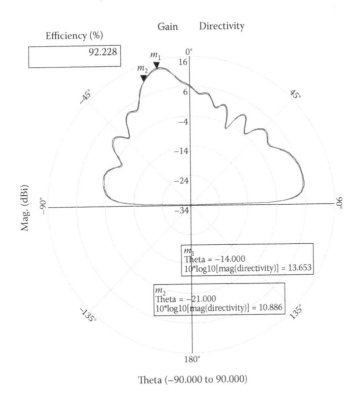

Gain   Directivity

Efficiency (%)

92.228

m_1

m_2

m_1
Theta = −14.000
10*log10[mag(directivity)] = 13.653

m_2
Theta = −21.000
10*log10[mag(directivity)] = 10.886

Theta (−90.000 to 90.000)

FIGURE 4.25 Radiation pattern of the stacked eight-element series fed array at 2.4 GHz.

4.10 STACKED SERIES PARALLEL FED MICROSTRIP 64-ELEMENT ARRAY

An S band, 2.4 GHz, stacked series parallel fed 64-element patch array was designed as shown in Figure 4.26. The resonator and the feed network were printed on a 1.6 mm thick substrate with relative dielectric constant of 2.25. The resonator is a rectangular patch resonator with dimensions of 38.2 × 38.2 mm. The distance between the array elements, d, is 96 mm. In the second layer the radiating element was printed on a 1.6 mm thick substrate with relative dielectric constant of 2.25. The electromagnetic field is coupled from the resonator to the radiating element. The radiating element is a square patch with dimensions $W = L = 40$ mm. The array consists of eight by eight elements. Each column in the array consists of an array with eight series fed elements. The eight series fed elements are combined by a parallel feed network. The parallel feed network is a 1:8 power combiner. The dimensions of the array aperture are 60 × 71.5 cm. The dimensions of the array with the feed network are around 60 × 78 cm. The array VSWR is better than 2:1 for 10% bandwidth. The array beam width is around 10° × 12°. The array directivity is around 24.1 dB. Losses in the feed network are around 1.5 dB. The antenna gain is around 22 dBi to 22.5 dBi.

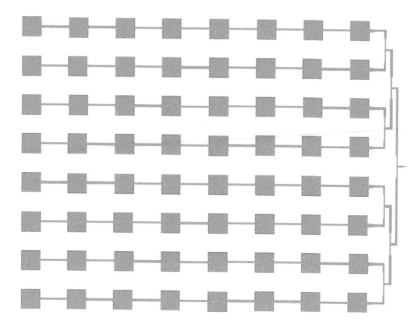

FIGURE 4.26   Sixty-four element stacked parallel series fed microstrip array.

4.11   CONCLUSIONS

Arrays function as a single antenna with a higher antenna gain than would be
obtained from the individual elements. In antenna arrays electromagnetic wave
interference is used to enhance the electromagnetic signal in one desired direction at
the expense of other directions. Microstrip antenna arrays possess attractive features
such as low profile, flexible, light weight, small volume, and low production cost. In
microstrip antenna arrays the feed network is an integral part of the antenna. The
feed network and radiating elements are printed on the same substrate. Gain limitations in microstrip antenna arrays due to losses in the feed network are the major
disadvantage of printed arrays. The efficiency of microstrip antenna arrays may be
improved significantly by reducing losses in the feed network. Efficient and compact
S, Ku, and Ka band arrays were presented in this chapter.

REFERENCES

1. Balanis, C. A. *Antenna Theory: Analysis and Design*, 2nd ed. New York: John Wiley &
   Sons, 1996.
2. James, J. R., Hall, P. S., and Wood, C. *Microstrip Antenna Theory and Design*. London:
   The Institution of Engineering and Technology, 1981.
3. Sabban, A., and Gupta, K. C. Characterization of Radiation Loss from Microstrip
   Discontinuities Using a Multiport Network Modeling Approach. *IEEE Transactions
   on Microwave Theory and Techniques*, 39(4):705–712, 1991.

4. Sabban, A. *Multiport Network Model for Evaluating Radiation Loss and Coupling Among Discontinuities in Microstrip Circuits.* PhD thesis, University of Colorado at Boulder, January 1991.
5. Kathei, P. B., and Alexopoulos, N. G. Frequency-Dependent Characteristic of Microstrip, Discontinuities in Millimeter-Wave Integrated Circuits. *IEEE Transactions on Microwave Theory and Techniques*, MTT-33, 33:1029–1035, 1985.
6. Sabban, A. A New Wideband Stacked Microstrip Antenna. In *IEEE Antenna and Propagation Symposium*, Houston, TX, June 1983.
7. Sabban, A., and Navon, E. A MM-Waves Microstrip Antenna Array. In *IEEE Symposium*, Tel Aviv, Israel, March 1983.
8. Sabban, A. Wideband Microstrip Antenna Arrays. In *IEEE Antenna and Propagation Symposium MELCOM*, Tel Aviv, Israel, June 1981.
9. Sabban, A. Microstrip Antennas. In *IEEE Symposium*, Tel Aviv, Israel, October 1979.

5 Applications of Low-Visibility Printed Antennas

5.1 INTRODUCTION

Microstrip antennas possess attractive features such as low profile, flexibility, light weight, small volume, and low production cost. In addition, the benefit of a compact low-cost feed network is attained by integrating the radio frequency (RF) feed network with the radiating elements on the same substrate. Microstrip antennas have been widely presented in books and papers in the last decade [1–8]. Microstrip antennas may be employed in communication links, seekers, and in biomedical systems.

In this chapter we present several applications of microstrip antennas. The design of mm wave microstrip antenna arrays with high efficiency is presented in this chapter. Gain limitation in microstrip antenna arrays due to losses in the feed network are presented in Ref. [1]. However, this discussion is limited to a 12-GHz plane slot array and radiation and dielectric losses are neglected.

The efficiency of microstrip antenna arrays may be improved significantly by reducing losses in the feed network. Losses in the microstrip feed network are due to conductor loss, radiation loss, and dielectric loss. Equations to calculate conductor loss and dielectric loss in microstrip lines are given in Ref. [1]. In Refs. [2,3] a planar multiport network modeling approach has been used to evaluate radiation loss from microstrip discontinuities. Full-wave analysis has been used by Kathei and Alexopoulos [4] to calculate radiation conductance of an open-circuited microstrip line. Full-wave analysis methods and software are available for characterization of microstrip discontinuities. These analyses include radiation effects also. The power radiated from a discontinuity may be evaluated from the computed S parameters. However, evaluation of power radiated from the computed S parameters requires a high numerical accuracy of the computed results. Therefore radiation loss values based on full-wave analysis are not widely available.

In this chapter losses in 64- and 256-patch antenna arrays at Ka band are evaluated. Methods to minimize these losses and to improve the antenna efficiency are presented. In the literature several applications of mm wave microstrip antenna arrays are described. Gain limitation in microstrip antenna arrays may be solved by employing active microstrip arrays.

However, active microstrip arrays have several disadvantages such as a significant increase in power consumption, weight, and dimensions. In some applications reflect arrays are employed. In reflect arrays the received power is transmitted by the same antenna array. The efficiency of microstrip antenna arrays may be improved by using a waveguide feed network. However, this results in a significant increase in the antenna weight and dimensions. In this case a transition from microstrip to waveguide is required.

Several imaging approaches are presented [9–14]. The common approach is based on an array of radiators (antennas) that receives radiation from a specific direction by using a combination of electronic and mechanical scanning. Another approach is based on a steering array of radiation sensors at the focal plane of a lens of a reflector. The sensor can be an antenna coupled to a resistor. In this chapter we present the development of a mm wave radiation detection array. The detection array may employ around 256- to 1024-patch antennas. These patches are coupled to a resistor. Optimization of the antenna structure, feed network dimensions, and resistor structure allows us to maximize the power rate dissipated on the resistor. Design considerations of the detection antenna array are given in this chapter.

Microstrip and printed antennas features make them excellent candidates to serve as antennas in biomedical systems. However, the antenna electrical performance is altered significantly in the vicinity of the human body. These facts significantly complicate the antenna design. The electrical performance of a new class of wideband wearable printed antennas for medical applications is presented in this chapter. RF transmission properties of human tissues have been investigated in several papers [15,16]. However, the effect of the human body on the electrical performance of the antennas at frequencies at which the biomedical system operates is not considered. The interaction between microstrip antennas and the human body is discussed in this chapter. The antenna bandwidth is around 10% for voltage standing wave ratio (VSWR) better than 2:1. The antenna beam width is around 100°. The antenna gain is around 4 dBi. If the air spacing between the sensors and the human body is increased from 0 mm to 5 mm the antenna resonant frequency is shifted by 5%.

5.2   LOW-VISIBILITY MICROSTRIP ANTENNA ARRAYS WITH HIGH EFFICIENCY

One of the major advantages of microstrip antennas is the simplicity of array construction [1]. The radiating elements may be etched jointly with the feed network as an integrated structure, leading to a very compact and low cost design. Although the technique for designing feed networks is well established, several difficulties are encountered while implementing it at mm wave frequencies. Microstrip line losses increase considerably at mm wave frequencies. Conductor, dielectric, and radiation losses are the major components of loss in mm wave microstrip antenna arrays. At frequencies ranging from 30 to 40 GHz, conductor losses are around 0.15 to 0.2 dB per wavelength; dielectric losses are around 0.045 dB per wavelength for a 50 Ω line on a 10 mil substrate with $\varepsilon_r = 2.2$.

The open nature of a microstrip configuration suffers from radiation. In mm wave microstrip antenna arrays, more bends, T-junctions, and other discontinuities are introduced in the feed network and radiation loss increases considerably. The multiport network model is employed to evaluate radiation loss from microstrip discontinuities. Minimization of losses in the microstrip feed network may result in microstrip antenna arrays with high efficiency.

5.2.1 Evaluation of Microstrip Feed Network Losses

Equations to calculate conductor loss and dielectric loss in microstrip lines are given in Ref. [1]. Dielectric loss is incorporated in the multiport network analysis by considering a complex dielectric constant. Conductor losses are included in the analysis by defining an equivalent loss tangent δ_c, given by $\delta_c = \delta_s/h$, where $\delta_s = \sqrt{\dfrac{2}{\omega\mu\sigma}}$. Where σ is the strip conductivity, h is the substrate height and μ is the free space permeability.

5.2.2 Evaluation of Radiation Loss

The multiport network model is employed to evaluate radiation loss from microstrip feed networks by adding a number of open ports at the edges of the planar model for the discontinuity structure. The multiport network model is based on the parallel-plate waveguide model [17] for microstrip lines. A similar network modeling approach has been used for analysis of microstrip discontinuities [2]. The planar waveguide model consists of two parallel conductors bounded by magnetic walls in the transverse directions. In this modeling approach for microstrip structures, fields underneath the microstrip configuration, the external fields (radiated fields, surface waves) are modeled separately in terms of multiport subnetworks by adding an equivalent edge admittance network connected to the edges of the microstrip configuration. These subnetworks are characterized in terms of Z-matrices that are evaluated by using Green's function approach. The subnetworks are combined using the segmentation technique to obtain circuit characteristics such as scattering parameters. Equations to compute the subnetworks Z-matrices are given in Ref. [3].

   The multiport network model is used to evaluate the voltage distributions at the open ports. Voltages at the discontinuity edges are represented by equivalent magnetic current sources, as shown in Figure 5.1. The amplitude M of the magnetic

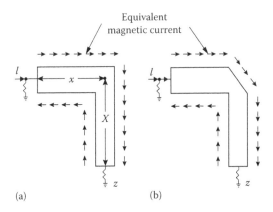

FIGURE 5.1 Equivalent magnetic current distribution at discontinuity edges. (a) Presents the equivalent magnetic current distribution at the edges of a right angle bend. (b) Presents the equivalent magnetic current distribution at the edges of a chamfered bend.

current elements is twice that of the edge voltage at that location, and the phase of the magnetic current is equal to the phase of the corresponding voltage. The total radiation is computed using the superposition of the far field radiated by each section. Referring to the coordinate system, shown in Figure 5.2, the far-field pattern $F(\theta, \phi)$ may be written in terms of voltages at the various elements.

With the voltage at the ith element as $(V(i) \, e^{j\alpha(i)})$, we have

$$F(\theta,\phi) = \sum_{i=1}^{N} 2V(i)W(i)\exp\{k_0\gamma_0(i)+\alpha(i)\}F_i(\theta,\phi) \qquad (5.1)$$

where

$$F_i(\theta,\phi) = \frac{\sin\left(\dfrac{k_0 W(i)}{2}\cos\theta\right)}{\dfrac{k_0 W(i)}{2}\cos\theta}\sin\theta$$

$$\gamma_0(i) = X_0(i)\sin\theta\cos\phi + Y_0(i)\cos\theta$$

where N is the number of ports, $X_0(i)$, $Y_0(i)$ specify the location of the ith magnetic current element, k_0 is the free space wave number, and $W(i)$ is the width of the ith element. The factor 2 in Equation 5.1 accounts for the image of the magnetic current

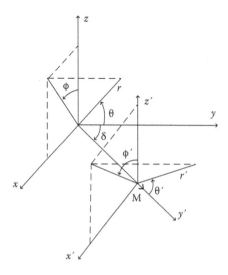

FIGURE 5.2 Coordinate system for external field calculations.

with respect to the ground plane. The radiated power is calculated by the integration of the Poynting vector over the half-space and may be written as

$$P_r = \frac{1}{240\pi} \int_{-\frac{\pi}{2}}^{\frac{\pi}{2}} \int_{0}^{\pi} \left(|E_\theta|^2 + |E_\phi|^2 \right) r^2 \sin\theta \, d\theta \, d\phi \tag{5.2}$$

The fields E_θ and E_ϕ are expressed in terms of $F(\theta, \phi)$ as

$$E_\theta = \hat{a}\theta \left(\frac{-jk_0}{4\pi r} F(\theta,\phi) F_\theta \right) \tag{5.3}$$

$$E_\phi = \hat{a}\phi \left(\frac{-jk_0}{4\pi r} F(\theta,\phi) F_\phi \right) \tag{5.4}$$

where

$$F_\phi = \sin\phi' \sin\phi + \cos\delta \cos\phi \cos\phi' \tag{5.5}$$

$$F_\theta = \sin\phi' \cos\theta \cos\phi + \cos\delta \cos\theta \sin\phi + \sin\delta \cos\phi' \sin\theta \tag{5.6}$$

$$\cos\theta' = \sin\theta \sin\phi \sin\delta + \cos\theta \cos\delta \tag{5.7}$$

$$\cos\phi' = \sin\theta \cos\phi / \sqrt{1 - \cos^2\theta'} \tag{5.8}$$

The radiated power may be expressed as a fraction of input power $10 \log_{10}(P_r/P_i)$ dB.

The radiation loss may be expressed as radiation loss (dB) = $10 \log_{10}(1 - P_r/P_i)$, where P_i is the input power at port 1. P_i is calculated from input current and the input impedance of the discontinuity terminated in matched loads at other ports.

5.2.3 RADIATION LOSS FROM MICROSTRIP DISCONTINUITIES

Computation of microstrip feed network losses are given in Refs. [2,3]. As an example, radiation loss of a right-angled bend in a 50 Ω line on a 10-mil Duroid® substrate with $\varepsilon_r = 2.2$ is 0.1 dB at 30 GHz and 0.17 dB at 40 GHz. Dielectric loss can be reduced by using substrates with a low dielectric loss. To minimize radiation loss, the number of discontinuities, such as bends and T-junctions, should be made as small as possible. Radiation from a curved microstrip line is much smaller compared to radiation from a right-angled bend. Moreover, to reduce radiation loss in the feed network, the width of the microstrip line is designed to be less than 0.12 λ on a 0.25 mm substrate with $\varepsilon_r = 2.2$.

Conductor loss may be minimized by designing the feed network length per wavelength as short as possible. By using a multilayer feed network design, the feed network length per wavelength is minimized considerably. Gold plating of the microstrip lines decreases conductor losses. Low loss coaxial cables may replace in the microstrip feed network long sections of microstrip lines. As an example, the insertion loss of a flexible cable at 30 GHz is 0.039 dB per centimeter. However, the insertion loss at 30 GHz of a 50 Ω microstrip line on a 10-mil substrate with $\varepsilon_r = 2.2$ is around 0.2 dB. By replacing a microstrip line with a length of 10 centimeters with a coaxial transmission line, the loss decreases by 1.6 dB. The coaxial line may be an integral part of the feed network or it may be embedded in the metallic ground plane. The transition from the microstrip line to the coaxial line is straightforward. The coaxial line center conductor is soldered to the microstrip line and the outer conductor is soldered or glued to the ground plane, by using conductive glue.

5.2.4   64- AND 256-MICROSTRIP ANTENNA ARRAYS WITH HIGH EFFICIENCY

Microstrip antenna arrays with integral feed networks may be broadly divided into arrays fed by parallel feeds and series fed arrays. Usually series fed arrays are more efficient than parallel fed arrays. However, parallel fed arrays have a well-controlled aperture distribution.

Two Ka band microstrip antenna arrays that consist of 64 radiating elements have been designed. The first array uses a parallel feed network and the second uses a parallel-series feed network as shown in Figure 5.3a and b. The performances of the arrays are compared in Table 5.1. Results given in Table 5.1 verify that the parallel series fed array is more efficient than the parallel fed array because of minimization of the number of discontinuities in the parallel-series feed network.

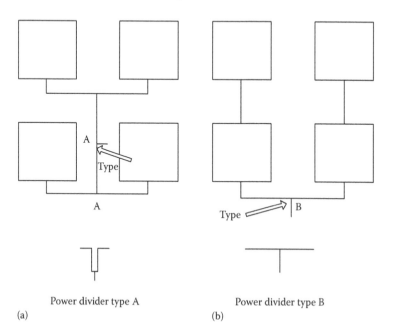

FIGURE 5.3   Configuration of 64-element microstrip antenna array. (a) Parallel feed network. (b) Parallel series feed network.

TABLE 5.1
Performance of 64-Element Microstrip Antenna Arrays

Parameter	Corporate Feed	Parallel Feed	Microstrip and Coaxial Corporate Feed
Number of elements	64	64	64
Beam width (°)	8.5	8.5	8.5
Computed gain (dBi)	26.3	26.3	26.3
Microstrip line loss (dB)	1.1	1.2	0.5
Radiation loss T-junction (dB)	0.27	0.45	0.27
Radiation loss bends (dB)	0.4	0.8	0.4
Radiation loss steps (dB)	0.045	–	0.045
Mismatch loss (dB)	0.5	0.5	0.5
Expected gain (dBi)	24.0	23.35	24.6
Efficiency (%)	58.9	50.7	67.6

The parallel-series fed array has been modified by using a 5-centimeter coaxial line to replace the same length of microstrip line. Results given in Table 5.1 indicate that the efficiency of the parallel series fed array that incorporates a coaxial line in the feed network is around 67.6% because of minimization of the microstrip line length.

Two microstrip antenna arrays that consist of 256 radiating elements have been designed. In the first array, type A, as shown in Figure 5.4a, the number of microstrip

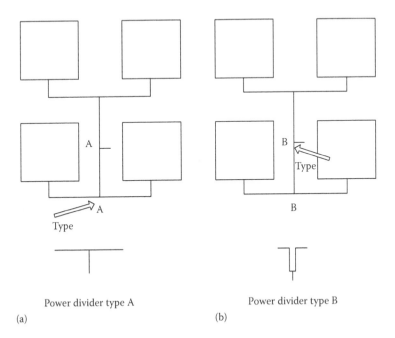

(a)   (b)

FIGURE 5.4 Configuration of 256-element microstrip antenna array. (a) Array type A. (b) Array type B.

TABLE 5.2

Performance of 256-Element Microstrip Antenna Arrays

Parameter	Type A	Type B	Type A and Microstrip Coaxial Feed
Number of elements	256	256	256
Beam width (°)	4.2	4.2	4.2
Computed gain (dBi)	32	32	32
Microstrip line loss (dB)	3.1	3.1	1.5
Radiation loss T-junction (dB)	0.72	0.72	0.72
Radiation loss bends (dB)	0.13	1.17	0.13
Radiation loss steps (dB)	0.12	–	0.12
Mismatch loss (dB)	0.5	0.5	0.5
Expected gain (dBi)	27.43	26.5	29.03
Efficiency (%)	34.9	28.2	50.47

discontinuities is minimized. The second array, type B, as shown in Figure 5.4b, incorporates more bend discontinuities in the feeding network. The performances of the arrays are compared in Table 5.2. The type A array with 256 radiating elements has been modified by using a 10-centimeter coaxial line to replace the same length of microstrip line. The array performances are compared in Table 5.2. Table 5.2 shows that the gain of the modified array has been increased by 1.6 dB. Results given in Table 5.2 verify that the type A array is more efficient than the type B array because of minimization of the number of bend discontinuities in the type A array feed network. The measured gain results are very close to the computed gain results and verify the loss computation presented in this chapter.

5.3 W BAND MICROSTRIP ANTENNA DETECTION ARRAY

Losses in the microstrip feed network are very high in the W band frequency range. In W band frequencies we may design a detection array. The array concept is based on an antenna coupled to a resistor. A direct antenna-coupling surface to a micro machined micro bridge resistor is used for heating and sensing. An analog complementary metal oxide semiconductor (CMOS) readout circuit may be employed as a sensing channel per pixel. Figure 5.5 presents a pixel block diagram.

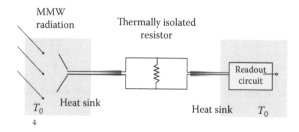

FIGURE 5.5   Antenna coupled to a resistor.

5.3.1 THE ARRAY PRINCIPLE OF OPERATION

The antenna receives effective mm wave radiation. The radiation power is transmitted to a thermally isolated resistor coupled to a titanium resistor. The electrical power raises the structure temperature with a short response time. The same resistor changes its temperature and therefore its electrical resistance. Figure 5.6 shows a single-array pixel. The pixel consists of a patch antenna, a matching network, printed resistor, and DC pads. The printed resistor consists of titanium lines and a titanium resistor coupled to an isolated resistor.

The operating frequency range of 92–100 GHz is the best choice. In the frequency range of 30–150 GHz there is a proven contrast between land, sky, and high transmittance of clothes. Size and resolution considerations promote higher frequencies above 100 GHz. Typical penetration of clothing at 100 GHz is 1 dB and 5–10 dB at 1 THz. Characterization and measurement considerations promote lower frequencies. The frequency range of 100 GHz allows sufficient bandwidth when working with illumination. The frequency range of 100 GHz is the best compromise. Figure 5.7 presents the array concept. Several types of printed antennas may be employed as the array element such as bowtie dipole, patch antenna, and ring resonant slot.

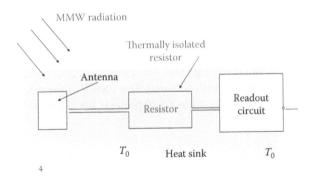

FIGURE 5.6   A single-array pixel.

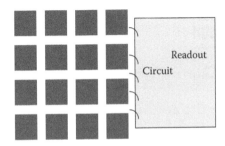

FIGURE 5.7   Array concept.

5.3.2   W Band Antenna Design

The bowtie dipole and a patch antenna have been considered as the array element. Computed results show that the directivity of the bowtie dipole is around 5.3 dBi and the directivity of a patch antenna is around 4.8 dBi. However, the length of the bowtie dipole is around 1.5 mm and the size of the patch antenna is around 700 × 700 μm. We used a quartz substrate with a thickness of 250 μm. The bandwidth of the bowtie dipole is wider than that of a patch antenna. However, the patch antenna bandwidth meets the detection array electrical specifications. We chose the patch antenna as the array element because the size of the patch is significantly smaller than that of the bowtie dipole. This feature allows us to design an array with a higher number of radiating elements. The resolution of a detection array with a higher number of radiating elements is improved. We also realized that the matching network between the antenna and the resistor has a smaller size for a patch antenna than for a bowtie dipole. The matching network between the antenna and the resistor consists of microstrip open stubs. The feed network determines the antenna efficiency. The insertion loss of a gold microstrip line with width of 1 μm and 188 μm length is 4.4 dB at 95 GHz. The insertion loss of a gold microstrip line with width of 10 μm and 188 μm length is 3.6 dB at 95 GHz. The insertion loss of a gold microstrip line with width of 20 μm and 188 μm length is 3.2 dB at 95 GHz. To minimize losses the feed line dimensions was selected as 60 × 10 × 1 μm. A taper connects the 10 μm width patch feed line to the 1 μm width Titanium resistor line. The resistor is thermally isolated from the patch antenna by using a 3 μm sacrificial layer. The patch is matched to the 10 μm width patch feed line by employing open circuited stubs. Figure 5.8 shows the 3D radiation pattern of the Bowtie dipole. Figure 5.9 presents the S_{11} parameter of the patch antenna. The electrical performance of the Bowtie dipole and the patch antenna was compared. The VSWR of the patch antenna is better than 2:1 for 10% bandwidth. Figure 5.10 presents the 3D radiation pattern of the patch antenna at 95 GHz.

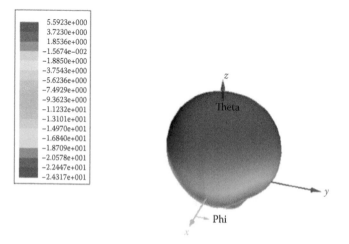

FIGURE 5.8   Dipole 3D radiation pattern.

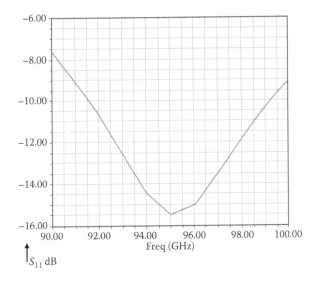

FIGURE 5.9 Patch S_{11} computed results.

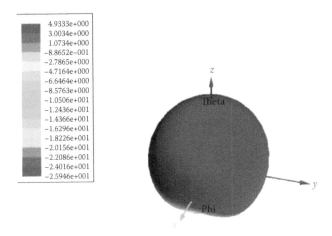

FIGURE 5.10 Patch 3D radiation pattern.

5.3.3 RESISTOR DESIGN

As described by Milkov [8], the resistor is thermally isolated from the patch antenna by using a sacrificial layer. Optimizations of the resistor structure maximize the power rate dissipated on the resistor. Ansoft HFSS software is employed to optimize the height of the sacrificial layer and the transmission line width and length. The dissipated power on a titanium resistor is higher than the dissipated power on a platinum resistor. The rate of the dissipated power on the resistor is around 25%. Material properties are given in Table 5.3. The sacrificial layer thickness may be 2–3 μm. Figure 5.11 shows the resistor configuration.

TABLE 5.3
Material Properties

Property	Units	siNi	Ti
Conductivity (K)	W/m/K	1.6	7
Capacity (C)	J/Kg/K	770	520
Density (ρ)	Gr/cm³	2.85	4.5
Resistance	Ω/□	>1e8	90
Thickness	μm	0.1	0.1

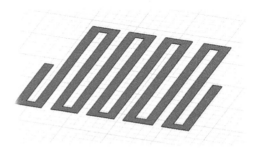

FIGURE 5.11   Resistor configuration.

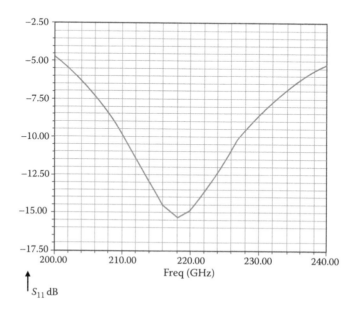

S_{11} dB

FIGURE 5.12   220-GHz patch S_{11} computed results.

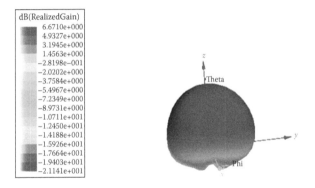

dB(RealizedGain)
6.6710e+000
4.9327e+000
3.1945e+000
1.4563e+000
−2.8198e−001
−2.0202e+000
−3.7584e+000
−5.4967e+000
−7.2349e+000
−8.9731e+000
−1.0711e+001
−1.2450e+001
−1.4188e+001
−1.5926e+001
−1.7664e+001
−1.9403e+001
−2.1141e+001

FIGURE 5.13   220-GHz patch 3D radiation pattern.

5.3.4   220-GHz Microstrip Patch Antenna

A quartz substrate with a thickness of 50–100 μm has been used to fabricate microstrip antennas at frequencies higher than 200 GHz. The size of the patch antenna is around 300×300 μm. Figure 5.12 presents the S_{11} parameters of the patch antenna. Figure 5.13 shows the 3D radiation pattern of the patch antenna.

5.4   MEDICAL APPLICATIONS OF MICROSTRIP ANTENNAS

Microstrip antennas possess attractive features such as low profile, flexibility, light weight, small volume, and low production cost. Microstrip and printed antenna features make them excellent candidates to serve as antennas in biomedical systems. However, the electrical performance of the antenna is altered significantly in the vicinity of the human body. These facts significantly complicate the antenna design. The electrical performance of a new class of wideband wearable printed antennas for medical applications is presented in this chapter. RF transmission properties of human tissues have been investigated in several papers [16,17]. However, the effect of the human body on the electrical performance of the antennas at frequencies at which biomedical systems operate is not presented. A new class of wideband compact wearable printed and microstrip antennas for medical applications is presented in this chapter.

5.4.1   Dual Polarized 434-MHz Printed Antenna

A new compact microstrip-loaded dipole antenna has been designed to provide horizontal polarization. The antenna consists of two layers. The first layer consists of a 0.8-mm RO3035 dielectric substrate. The second layer consists of a 0.8 mm RT/Duroid 5880 dielectric substrate. The substrate thickness determines the antenna bandwidth. However, with a thinner substrate we may achieve better flexibility. We also designed a thicker double-layer microstrip-loaded dipole antenna with wider bandwidth. A printed slot antenna provides a vertical polarization. The proposed antenna is dual polarized. The printed dipole and the slot antenna provide dual orthogonal polarizations. The dual polarized antenna is shown in Figure 5.14. The

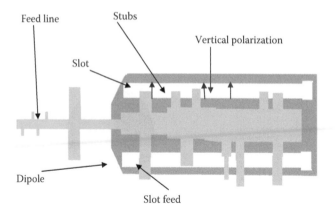

FIGURE 5.14 Printed dual polarized antenna.

antenna dimensions are $26 \times 6 \times 0.16$ cm. The antenna may be employed as a wearable antenna on a human body. The antenna may be attached to the patient's shirt in the patient's stomach zone. Alternatively, the antenna may be attached to the patient's back. The antenna has been analyzed using Agilent ADS software. There is a good agreement between measured and computed results.

The antenna bandwidth is around 10% for VSWR better than 2:1, as shown in Figure 5.15. The antenna beam width is around 100°. The antenna gain is around 4 dBi. The computed S_{11} and S_{22} parameters are shown in Figure 5.15. Figure 5.16 presents the antenna measured S_{11} parameters. The computed radiation pattern is shown in Figure 5.17. The antenna cross-polarized field strength may be adjusted by varying the slot feed location. The antenna dimensions may be reduced to $6 \times 6 \times 0.16$ cm by employing a folded antenna configuration as shown in Figure 5.18. Figure 5.19 presents the antenna computed S_{11} and S_{22} parameters. The computed radiation pattern of the folded dipole is shown in Figure 5.20.

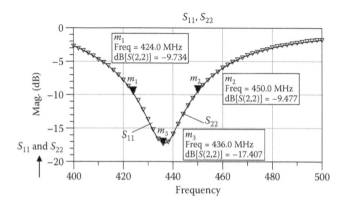

FIGURE 5.15 Computed S_{11} and S_{22} results.

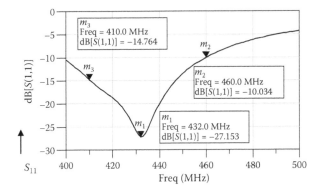

FIGURE 5.16   Measured S_{11} on the human body.

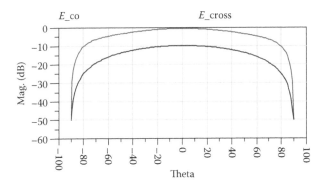

FIGURE 5.17   Antenna radiation pattern.

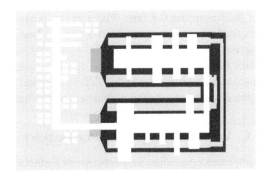

FIGURE 5.18   Folded dual polarized antenna.

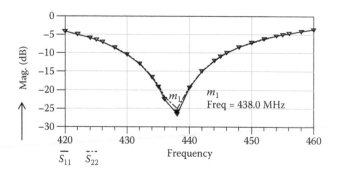

FIGURE 5.19 Folded antenna computed S_{11} and S_{22} results.

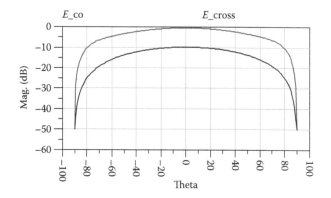

FIGURE 5.20 Folded antenna radiation pattern.

5.4.2 NEW LOOP ANTENNA WITH A GROUND PLANE

A new loop antenna with a ground plane has been designed on Kapton substrates with thicknesses of 0.25 mm and 0.4 mm. The antenna is shown in Figure 5.21. Figure 5.22 presents the loop antenna computed S_{11} on the human body. The computed radiation pattern is shown in Figure 5.23.

Table 5.4 compares the electrical performance of a loop antenna with a ground plane with a loop antenna without a ground plane. There is a good agreement between measured and computed results of antenna parameters on the human body.

5.4.3 ANTENNA S_{11} VARIATION AS A FUNCTION OF DISTANCE FROM THE BODY

The antenna input impedance variation as a function of distance from the body has been computed by employing ADS software. The analyzed structure is presented in Figure 5.24. Properties of human body tissues are listed in Table 5.5 (see Ref. [16]). These

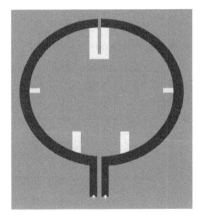

FIGURE 5.21 Loop antenna with a ground plane.

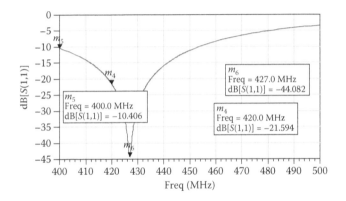

FIGURE 5.22 Computed S_{11} of loop antenna.

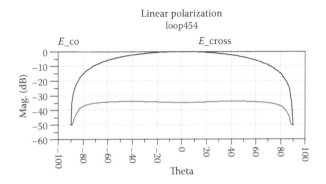

FIGURE 5.23 Loop antenna radiation pattern on the human body.

TABLE 5.4

Comparison of Loop Antennas

Antenna	Beam Width (3 dB)	Gain (dBi)	VSWR
Loop no ground plane	100°	0	4:1
Loop with ground plane	100°	0	2:1

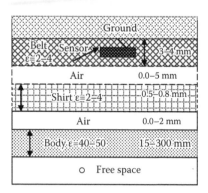

FIGURE 5.24   Analyzed structure for impedance calculations.

TABLE 5.5

Properties of Human Body Tissues

Tissue	Property	434 MHz	600 MHz
Skin	σ	0.57	0.6
	ε	41.6	40.43
Stomach	σ	0.67	0.73
	ε	42.9	41.41
Colon, muscle	σ	0.98	1.06
	ε	63.6	61.9
Lung	σ	0.27	0.27
	ε	38.4	38.4

properties were employed in the antenna design. Figure 5.25 presents S_{11} results for different belt thicknesses, shirt thicknesses, and air spacing between the antennas and the human body. One may conclude from results shown in Figure 5.26 that the antenna has a VSWR better than 2.5:1 for air spacing up to 8 mm between the antennas and the patient's body. For frequencies ranging from 415 MHz to 445 MHz the antenna is matched when there is no air spacing between the antenna and the patient's body. Figure 5.27 presents S_{11} folded antenna results for different positions

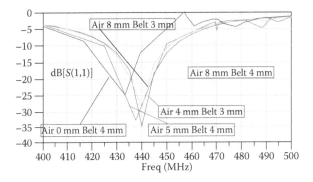

FIGURE 5.25 S_{11} results for different antenna positions relative to the human body.

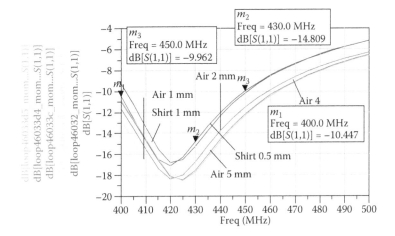

FIGURE 5.26 Folded antenna S_{11} results for different positions relative to the human body.

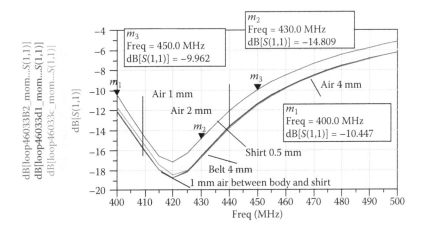

FIGURE 5.27 S_{11} results for different belt thicknesses.

TABLE 5.6
Explanation of Figure 5.26

Plot Color	Sensor Position
Red	Shirt thickness 0.5 mm
Blue	Shirt thickness 1 mm
Pink	Air spacing 2 mm
Green	Air spacing 4 mm
Sky	Air spacing 1 mm
Purple	Air spacing 5 mm

TABLE 5.7
Explanation of Figure 5.27

Color	Sensor Position
Red	Shirt thickness 0.5 mm
Blue	Air spacing body to shirt 1 mm
Pink	Belt thickness 4 mm
Sky	Air spacing shirt to belt 1 mm
Green	Air spacing shirt to belt 4 mm

relative to the human body. An explanation of Figure 5.26 is given in Table 5.6. From results shown in Figure 5.13 we can see that the folded antenna has a VSWR better than 2.0:1 for air spacing up to 5 mm between the antennas and the patient's body. If the air spacing between the sensors and the human body is increased from 0 mm to 5 mm the antenna resonant frequency is shifted by 5%.

S_{11} results in Figure 5.27 present the folded antenna matching when the belt thickness was changed from 2 to 4 mm. An explanation of Figure 5.27 is given in Table 5.7. S_{11} results are better than -10 dB for belt thicknesses ranging from 2 to 4 mm. Computed S_{11} and S_{22} results were better than -10 dB for different body tissues with dielectric constants ranging from 40 to 50. Computed S_{11} and S_{22} results were better than -10 dB for different shirts and belts with dielectric constants ranging from 2 to 4.

5.4.4 MEDICAL APPLICATIONS FOR LOW-VISIBILITY ANTENNAS

An application of the proposed antenna is shown in Figure 5.28. Three to four folded dipole antennas may be assembled in a belt and attached to the patient's stomach. The cable from each antenna is connected to a recorder. The received signal is routed to a switching matrix. The signal with the highest level is selected during the medical test. The antennas receive a signal that is transmitted from various positions in the human body. Folded antenna may be also attached on the patient's back to improve the level of the received signal from different locations in the human body. Figures 5.29 and 5.30 show various antenna locations on the back and

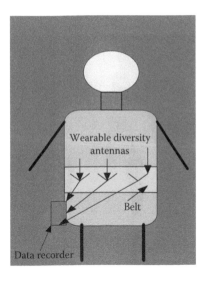

FIGURE 5.28 Wearable antenna.

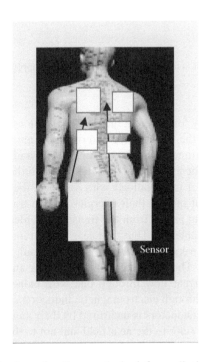

FIGURE 5.29 Printed antenna locations on the back for medical applications.

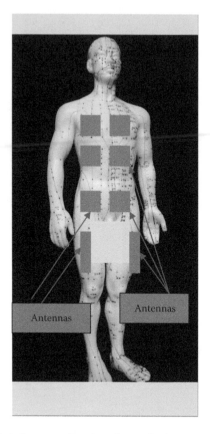

FIGURE 5.30 Printed pitch antenna locations for medical applications.

front of the human body for different medical applications. In several applications the distance separating the transmitting and receiving antennas is less than $2D^2/\lambda$, where D is the largest dimension of the source of the radiation. In these applications the amplitude of the electromagnetic field close to the antennas may be quite powerful, but because of rapid fall-off with distance, they do not radiate energy to infinite distances, but instead their energies remain trapped in the region near the antenna, not drawing power from the transmitter unless they excite a receiver in the area close to the antenna. Thus, the near fields transfer energy only to close distances from the receivers, and when they do, the result is felt as an extra power draw in the transmitter. The receiving and transmitting antennas are magnetically coupled. Change in current flow through one wire induces a voltage across the ends of the other wire through electromagnetic induction. The amount of inductive coupling between two conductors is measured by their mutual inductance. In these applications we have to refer to the near field and not to the far field radiation pattern. In Figures 5.31 and 5.32 several microstrip antennas for medical applications at 434 MHz are shown.

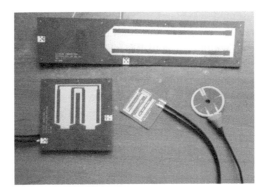

FIGURE 5.31 Microstrip antennas for medical applications.

FIGURE 5.32 Microstrip antennas for medical applications.

5.5 CONCLUSION

A 64-microstrip antenna array with efficiency of 67.6% and a 256-microstrip antenna array with efficiency of 50.47% have been presented in this chapter. Methods to reduce losses in mm wave microstrip antenna arrays have been described in this chapter. The results presented in this chapter point out that radiation losses need to be taken into account for accurate microstrip antenna array design at mm wave frequencies. By minimizing the number of bend discontinuities the gain of the 256-microstrip antenna array has been increased by 1 dB.

Several applications of mm wave microstrip antenna arrays have been presented. Losses in the microstrip feed network constitute a significant limit on the possible applications of microstrip antenna arrays in mm wave frequencies. MM wave microstrip antenna arrays may be employed in communication links, seekers, and detection arrays. The array may consist of around 256–1024 elements. Design considerations of the antenna and the feed network are given in this chapter. Optimization of the antenna structure and feed network allows us to design and fabricate microstrip antenna arrays with high efficiency.

This chapter presented wideband microstrip antennas with high efficiency for medical applications. The antenna bandwidth is around 10% for VSWR better than 2:1. The antenna beam width is around 100°. The antenna gain is around 2 dBi. The antenna S_{11} results for different belt thicknesses, shirt thicknesses, and air spacing between the antennas and the human body are given in this chapter. The effect of the antenna location on the human body should be considered in the antenna design process. If the air spacing between the sensors and the human body is increased from 0 mm to 5 mm the antenna resonant frequency is shifted by 5%. The proposed antenna may be employed in Medicare RF systems.

REFERENCES

1. James, J. R., Hall, P. S., and Wood, C. *Microstrip Antenna Theory and Design.* London: The Institution of Engineering and Technology, 1981.
2. Sabban, A., and Gupta, K. C. Characterization of Radiation Loss from Microstrip Discontinuities Using a Multiport Network Modeling Approach. *IEEE Transactions on Microwave Theory and Techniques,* 39(4):705–712, 1991.
3. Sabban, A. *Multiport Network Model for Evaluating Radiation Loss and Coupling Among Discontinuities in Microstrip Circuits.* PhD thesis, University of Colorado at Boulder, January 1991.
4. Kathei, P. B., and Alexopoulos, N. G. Frequency-Dependent Characteristic of Microstrip, Discontinuities in Millimeter-Wave Integrated Circuits. *IEEE Transactions on Microwave Theory and Techniques,* MTT-33, 1029–1035, 1985.
5. Sabban, A. A New Wideband Stacked Microstrip Antenna. In *IEEE Antenna and Propagation Symposium,* Houston, TX, June 1983.
6. Sabban, A. Microstrip Antennas. In *IEEE Symposium,* Tel Aviv, Israel, October 1979.
7. Sabban, A. Wideband Microstrip Antenna Arrays. In *IEEE Antenna and Propagation Symposium MELCOM,* Tel Aviv, Israel, June 1981.
8. Sabban, A., and Navon, E. A MM-Waves Microstrip Antenna Array. In *IEEE Symposium,* Tel Aviv, Israel, March 1983.
9. Milkov, M. M. *Millimeter-Wave Imaging System Based on Antenna-Coupled Bolometer.* MSc.Thesis, UCLA, 2000.
10. de Lange, G. et al. A 3*3 mm-Wave Micro Machined Imaging Array with Sis Mixers. *Applied Physics Letters,* 75(6):868–870, 1999.
11. Rahman, A. et al. Micromachined Room Temperature Microbolometers for mm-Wave Detection. *Applied Physics Letters,* 68(14):2020–2022, 1996.
12. Sinclair, G. N. et al. Passive Millimeter Wave Imaging in Security Scanning. *Proceedings SPIE,* 4032:40–45, 2000.
13. Luukanen, A. et al. US Patent 6242740, 2001.
14. Jack, M. D. et al. US Patent 6329655, 2001.
15. Chirwa, L. C., Hammond, P. A., Roy, S., and Cumming, D. R. S. Electromagnetic Radiation from Ingested Sources in the Human Intestine Between 150 MHz and 1.2 GHz. *IEEE Transactions on Biomedical Engineering,* 50(4):484–492, 2003.
16. Werber, D., Schwentner, A., and Biebl, E. M. Investigation of RF Transmission Properties of Human Tissues. *Advances in Radio Science,* 4:357–360, 2006.
17. Kompa, G., and Mehran, R. Planar Waveguide Model for Computing Microstrip Components. *Electronic Letters,* 11(9):459–460, 1975.

6 Wearable Antennas for Communication and Medical Applications

6.1 INTRODUCTION

Communication and biomedical industry is in continuous growth in the last few years. Low-profile compact antennas are crucial in the development of wearable human communication and biomedical systems. The proposed antennas may be linearly or dually polarized. Design considerations, computational results, and measured results on the human body of several compact wideband microstrip antennas with high efficiency at 434 MHz ± 5% are presented in this chapter. The compact dually polarized antenna dimensions are 5 × 5 × 0.05 cm. The antenna beam width is around 100°. The antenna gain is around 0–4 dBi. The proposed antenna may be used in communication and Medicare RF systems. The antenna S_{11} results for different belt thickness, shirt thickness, and air spacing between the antennas and the human body are presented in this chapter. If the air spacing between the new dually polarized antenna and the human body is increased from 0 to 5 mm, the antenna resonant frequency is shifted by about 5%. Varactors may be used to tune the antenna resonant frequency.

Microstrip antennas are widely employed in communication systems and seekers. Microstrip antennas posse's attractive features that are crucial to communication and medical systems. Microstrip antennas features are low profile, flexible, light weight, and have low production cost. In addition, the benefit of a compact low-cost feed network is attained by integrating the radio frequency (RF) front end with the radiating elements on the same substrate. Microstrip antennas have been widely presented in books and papers in the last decade [1–7]. However, the effect of the human body on the electrical performance of wearable antennas at 434 MHz is not presented [8–13]. RF transmission properties of human tissues have been investigated in several articles [8,9]. Several wearable antennas have been presented in the last decade [10–14]. A review of wearable and body-mounted antennas designed and developed for various applications at different frequency bands over the last decade can be found in Ref. [10]. In Ref. [11] meander wearable antennas in close proximity of a human body are presented in the frequency range between 800 MHz and 2700 MHz. In Ref. [12] a textile antenna performance in the vicinity of the human body is presented at 2.4 GHz. In Ref. [13] the effect of the human body on wearable 100-MHz portable radio antennas is studied. In Ref. [13] the authors concluded that wearable antennas need to be shorter by 15%–25% from the antenna length in free space. Measurement of the antenna gain in Ref. [13] shows that a wide dipole (116 × 10 cm) has a –13 dBi gain. The antennas presented in Refs. [10–13] were developed mostly for cellular applications. Requirements and the frequency range for medical applications are different from those for cellular applications.

In this chapter, a new class of wideband compact wearable microstrip antennas for medical applications is presented. Numerical results with and without the presence of the human body are discussed. The antennas' voltage standing wave ratio (VSWR) is better than 2:1 at 434 MHz ± 5%. The antenna beam width is around 100°. The antennas gain is around 0–4 dBi. The antenna resonant frequency is shifted by 5% if the air spacing between the antenna and the human body is increased from 0 mm to 5 mm.

6.2   DUALLY POLARIZED WEARABLE 434-MHz PRINTED ANTENNA

A compact microstrip-loaded dipole antenna has been designed to provide horizontal polarization. The antenna dimensions have been optimized to operate on the human body by employing Agilent Advanced Design System (ADS) software [15]. The antenna consists of two layers. The first layer consists of a 0.8 mm RO3035 dielectric substrate. The second layer consists of a 0.8 mm RT/Duroid® 5880 dielectric substrate. The substrate thickness determines the antenna bandwidth. However, thinner antennas are flexible. Thicker antennas have been designed with a wider bandwidth. The printed slot antenna provides a vertical polarization. In several medical systems the required polarization may be vertical or horizontal. The proposed antenna is dually polarized. The printed dipole and the slot antenna provide dual orthogonal polarizations. The dimensions and current distribution of the dual polarized wearable antenna are presented in Figure 6.1. The antenna dimensions are 26 × 6 × 0.16 cm. Figure 6.1 indicates that all antenna parts contribute to the antenna radiation pattern. The antenna may be used as a wearable antenna on a human body. The antenna may be attached to the patient's shirt, stomach, or in the back zone. The antenna has been analyzed by using Agilent ADS software. There is a good agreement between measured and computed results. The antenna bandwidth is around 10% for VSWR better than 2:1. The antenna beam width is around 100°. The antenna gain is around 2 dBi. The computed S_{11} and S_{22} parameters are presented in Figure 6.2. Figure 6.3

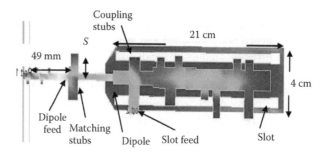

FIGURE 6.1   Current distribution of the wearable antenna.

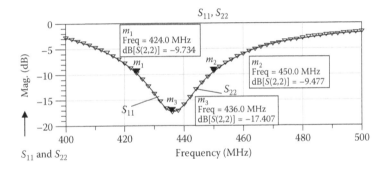

FIGURE 6.2 Computed S_{11} and S_{22} results on human body.

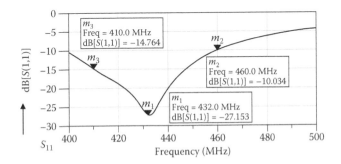

FIGURE 6.3 Measured S_{11} on the human body.

presents the antenna measured S_{11} parameters. The computed radiation patterns are shown in Figure 6.4. The copolar radiation pattern belongs to the y–z plane. The cross-polar radiation pattern belongs to the x–z plane. The antenna cross-polarized field strength may be adjusted by varying the slot feed location. The dimensions and current distribution of the folded dually polarized antenna presented in Figure 6.5. The antenna dimensions are $5.5 \times 4 \times 0.16$ cm. Figure 6.6 indicates that all antenna parts contribute to the antenna radiation pattern. The computed radiation patterns of the folded dipole are shown in Figure 6.7. The antenna radiation characteristics on

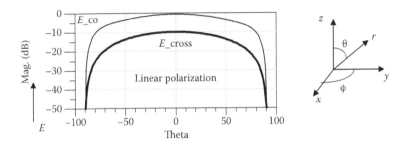

FIGURE 6.4 Antenna radiation patterns.

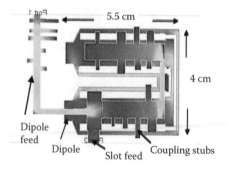

FIGURE 6.5   Current distribution of the folded dipole antenna, 7 × 5 × 0.16 cm.

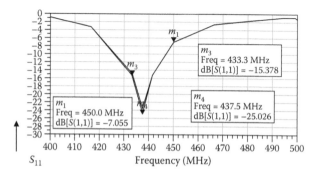

FIGURE 6.6   Folded antenna computed S_{11} and S_{22} results.

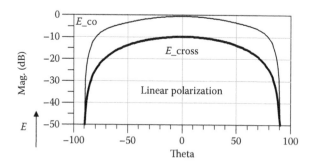

FIGURE 6.7   Folded antenna radiation patterns.

the human body have been measured by using a phantom. The phantom electrical characteristics represent the electrical characteristics of the human body.

The phantom has a cylindrical shape with a 40-cm diameter and a length of 1.5 m. The phantom contains a mix of 55% water, 44% sugar, and 1% salt. The antenna under test was placed on the phantom during the measurements of the antenna's radiation characteristics. S_{11} and S_{12} parameters were measured directly on the human body by using a network analyzer. The measured results were compared to measurements from a known reference antenna.

6.3   LOOP ANTENNA WITH GROUND PLANE

A new loop antenna with ground plane has been designed on Kapton substrates with thickness of 0.25 and 0.4 mm. The antenna without ground plane is shown in Figure 6.8. The loop antenna VSWR without the tuning capacitor was 4:1. This loop antenna may be tuned by adding a capacitor or varactor as shown in Figure 6.9. Tuning the antenna allows us to work in a wider bandwidth. Figure 6.10 presents the loop antenna computed S_{11} on human body. There is a good agreement between measured and computed S_{11}. The computed 3D radiation pattern is shown in Figure 6.11.

Table 6.1 compares the electrical performance of a loop antenna with ground plane to a loop antenna without ground plane. There is a good agreement between measured and computed results for several loop antenna electrical parameters on the human body. The results presented in Table 6.1 indicate that the loop antenna with

FIGURE 6.8   Loop antenna with ground plane printed on 0.25-mm-thick substrate.

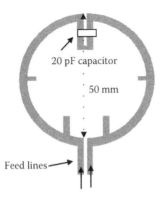

FIGURE 6.9   Tunable loop antenna without ground plane.

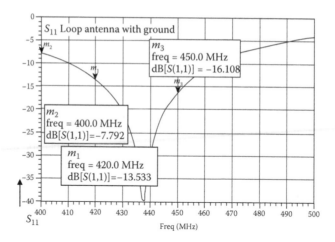

FIGURE 6.10 Computed S_{11} of new loop antenna, presented in Figure 6.8.

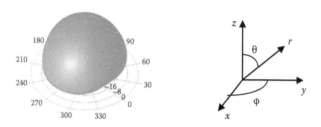

FIGURE 6.11 Loop antenna with ground, 3D radiation pattern.

TABLE 6.1

Electrical Performance of Several Loop Antenna Configurations

Antenna with No Tuning Capacitor	Beam Width 3 dB	Gain dBi	VSWR
Loop no GND	100°	0	4:1
Loop with tuning capacitor (no GND)	100°	0	2:1
Wearable loop with GND	100°	0–2	2:1
Loop with GND in free space	100° to 110°	−3	5:1

ground plane is matched to the human body environment, without the tuning capacitor, better than the loop antenna without ground plane.

The computed 3D radiation pattern and the coordinate used in this chapter are shown in Figure 6.11. The computed S_{11} of the loop antenna with a tuning capacitor is given in Figure 6.12. Figure 6.13 presents the radiation pattern of loop antenna without ground on the human body. Figure 6.14 presents a loop antenna with ground plane printed on a 0.4-mm-thick substrate. Figure 6.15 presents the loop antenna

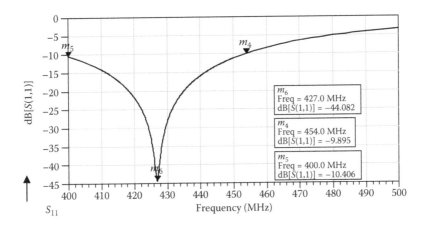

m_6
Freq = 427.0 MHz
dB[$S(1,1)$] = −44.082

m_4
Freq = 454.0 MHz
dB[$S(1,1)$] = −9.895

m_5
Freq = 400.0 MHz
dB[$S(1,1)$] = −10.406

S_{11}

FIGURE 6.12   Computed S_{11} of a tuned loop antenna, without ground plane.

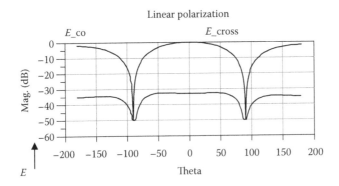

Linear polarization

*E*_co   *E*_cross

E

FIGURE 6.13   Radiation pattern of loop antenna without ground on human body.

FIGURE 6.14   Loop antenna with ground plane printed on 0.4 mm thick substrate.

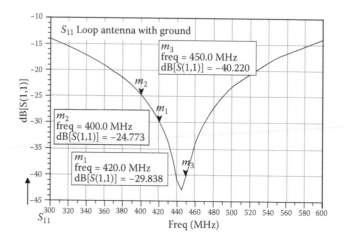

FIGURE 6.15 Computed S_{11} of new loop antenna printed on 0.4 mm thick substrate.

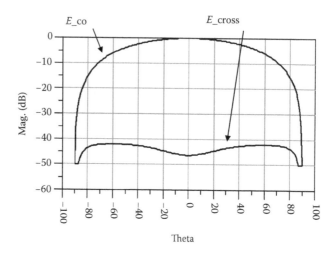

FIGURE 6.16 New loop antenna, printed on 0.4 mm thick substrate, radiation pattern.

computed S_{11} on the human body. Loop antennas printed on a thicker substrate have wider bandwidth as presented in Figure 6.15. Figure 6.16 presents the loop antenna, printed on a 0.4-mm-thick substrate, with a radiation pattern.

6.4 ANTENNA S_{11} VARIATION AS A FUNCTION OF DISTANCE FROM THE BODY

The antenna input impedance variation as function of distance from the body was computed by employing ADS software. The analyzed structure is presented in Figure 6.17a. The antenna was placed inside a belt with thickness between 1 to 4 mm as shown in Figure 6.17b. The patient body thickness was varied from 15 mm to 300 mm. The dielectric constant of the body was varied from 40 to 50. The antenna was placed inside a belt

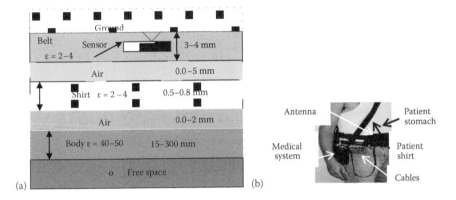

FIGURE 6.17   (a) Analyzed structure model. (b) Medical system on patient.

with thickness between 2 and 4 mm with dielectric constant from 2 to 4. The air layer between the belt and the patient's shirt may vary from 0 mm to 8 mm. The shirt thickness was varied from 0.5 mm to 1 mm. The dielectric constant of the shirt was varied from 2 to 4. Properties of human body tissues are listed in Table 6.2 (see Ref. [8]). These properties were employed in the antenna design. Figure 6.18 presents S_{11} results (of the antenna shown in Figure 6.1) for different belt thicknesses, shirt thicknesses, and air spacing between the antennas and human body.

One may conclude from results shown in Figure 6.18 that the antenna has VSWR better than 2.5:1 for air spacing up to 8 mm between the antennas and patient body. For frequencies ranging from 415 MHz to 445 MHz the antenna has VSWR better than 2:1 when there is no air spacing between the antenna and the patient body.

TABLE 6.2
Properties of Human Body Tissues

Tissue	Property	434 MHz	800 MHz	1000 MHz
Prostate	σ	0. 75	0.90	1.02
	ε	50.53	47.4	46.65
Stomach	σ	0.67	0.79	0.97
	ε	42.9	40.40	39.06
Colon, Heart	σ	0.98	1.15	1.28
	ε	63.6	60.74	59.96
Kidney	σ	0.88	0.88	0.88
	ε	117.43	117.43	117.43
Nerve	σ	0.49	0.58	0.63
	ε	35.71	33.68	33.15
Fat	σ	0.045	0.056	0.06
	ε	5.02	4.58	4.52
Lung	σ	0.27	0.27	0.27
	ε	38.4	38.4	38.4

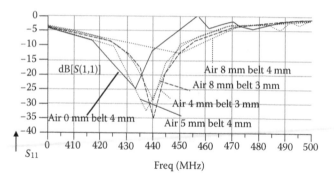

FIGURE 6.18 S_{11} results for different antenna positions relative to the human body.

Results shown in Figure 6.19 indicate that the folded antenna (the antenna shown in Figure 6.5) has VSWR better than 2.0:1 for air spacing up to 5 mm between the antennas and the patient body. Figure 6.19 presents S_{11} results of the folded antenna for different positions relative to the human body. Explanation of Figure 6.19 is given in Table 6.3. If the air spacing between the sensors and the human body is increased from 0 to 5 mm, the antenna resonant frequency is shifted by 5%. The loop antenna with ground plane has VSWR better than 2.0:1 for air spacing up to 5 mm between the antennas and the patient body. If the air spacing between the sensors and the human body is increased from 0 to 5 mm, the computed antenna resonant frequency is shifted by 2%. However, if the air spacing between the sensors and the human body is increased up to 5 mm, the measured loop antenna resonant frequency is shifted by 5%. Explanation of Figure 6.20 is given in Table 6.4.

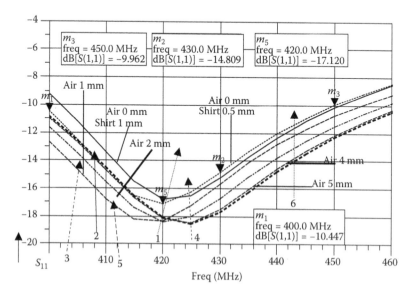

FIGURE 6.19   Folded antenna S_{11} results for different antenna position relative to the human body.

TABLE 6.3
Explanation of Figure 6.19

Picture #	Line Type	Sensor Position
1	Dot	Shirt thickness 0.5 mm
2	Line	Shirt thickness 1 mm
3	Dash dot	Air spacing 2 mm
4	Dash	Air spacing 4 mm
5	Long dash	Air spacing 1 mm
6	Big dots	Air spacing 5 mm

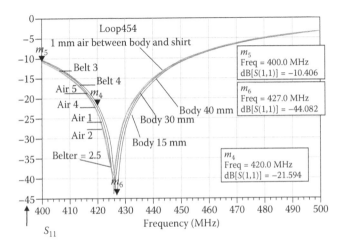

FIGURE 6.20   Loop antenna S_{11} results for different antenna positions relative to the body.

TABLE 6.4
Explanation of Figure 6.17

Plot Color	Sensor Position
Red	Body 15 mm, air spacing 0 mm
Blue	Body 15 mm, air spacing 5 mm
Pink	Body 40 mm, air spacing 0 mm
Green	Body 30 mm, air spacing 0 mm
Sky	Body 15 mm, air spacing 2 mm
Purple	Body 15 mm, air spacing 4 mm

6.5   WEARABLE ANTENNAS

An application of the proposed antenna is shown in Figure 6.21. Three to four folded dipole or loop antennas may be assembled in a belt and attached to the patient's stomach. The cable from each antenna is connected to a recorder. The received signal is routed to a switching matrix. The signal with the highest level is selected during the medical test. The antennas receive a signal that is transmitted from various positions in the human body. Folded antennas may be also attached on the patient's back to improve the level of the received signal from different locations in the human body. Figures 6.22 and 6.23 show various antenna locations on the back and front of the human body for different medical applications.

In several applications the distance separating the transmitting and receiving antennas is less than $2D^2/\lambda$. D is the largest dimension of the radiator. In these applications the amplitude of the electromagnetic field close to the antenna may be quite powerful, but because of rapid fall-off with distance, the antenna do not radiate energy to infinite distances, but instead the radiated power remains trapped in the region near the antenna. Thus, the near-fields transfer energy only to close distances from the receivers. The receiving and transmitting antennas are magnetically coupled. A change in current flow through one wire induces a voltage across the ends of the other wire through electromagnetic induction. The amount of inductive coupling between two conductors is measured by their mutual inductance. In these applications we have to refer to the near-field and not to the far-field radiation.

In Figures 6.23 and 6.24 several microstrip antennas for medical applications at 434 MHz are shown. The backside of the antennas is presented in Figure 6.23b. The diameter of the loop antenna presented in Figure 6.24 is 50 mm. The dimensions of the folded dipole antenna are $7 \times 6 \times 0.16$ cm. The dimensions of the compact folded dipole presented in Figure 6.24 are $5 \times 5 \times 0.5$ cm.

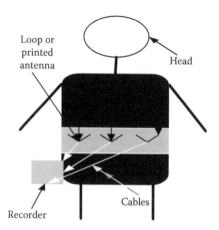

FIGURE 6.21   Printed wearable antenna.

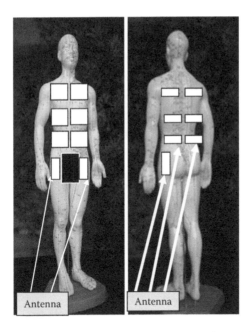

FIGURE 6.22   Printed patch antenna locations for various medical applications.

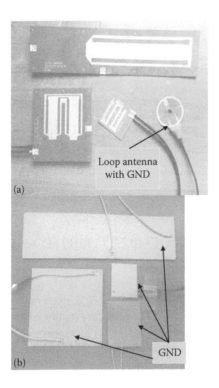

FIGURE 6.23   (a) Microstrip antennas for medical applications. (b) Backside of the antennas.

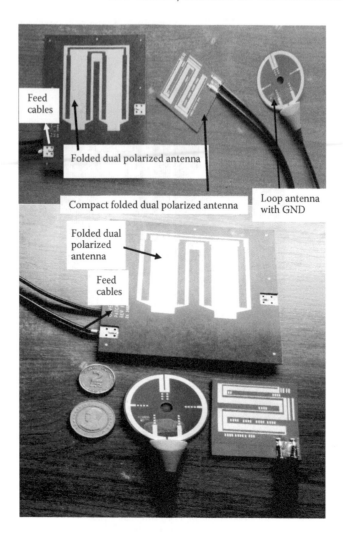

FIGURE 6.24 Microstrip antennas for medical applications.

6.6 COMPACT DUAL POLARIZED PRINTED ANTENNA

A new compact microstrip loaded dipole antennas has been designed. The antenna consists of two layers. The first layer consists of a 0.25 mm FR4 dielectric substrate. The second layer consists of a 0.25 mm Kapton dielectric substrate. The substrate thickness determines the antenna bandwidth. However, with a thinner substrate we may achieve better flexibility. The proposed antenna is dual polarized. The printed dipole and the slot antenna provide dual orthogonal polarizations. The dual polarized antenna is shown in Figure 6.25. The antenna dimensions are 5 × 5 × 0.05 cm.

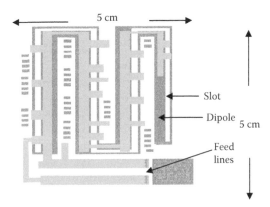

FIGURE 6.25 Printed compact dual polarized antenna.

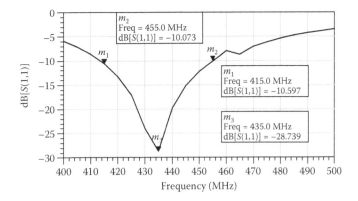

FIGURE 6.26 Computed S_{11} results of compact antenna.

The antenna may be attached to the patient's shirt in the patient's stomach or back zone. The antenna has been analyzed by using Agilent ADS software. There is a good agreement between measured and computed results. The antenna bandwidth is around 10% for VSWR better than 2:1. The antenna beam width is around 100°. The antenna gain is around 0 dBi. The computed S_{11} parameters are presented in Figure 6.26. Figure 6.27 presents the antenna measured S_{11} parameters. The antenna cross-polarized field strength may be adjusted by varying the slot feed location. The computed 3D radiation pattern of the antenna is shown in Figure 6.28. The computed radiation pattern is shown in Figure 6.29.

6.7 HELIX ANTENNA PERFORMANCE ON THE HUMAN BODY

To compare the variation of the new antenna input impedance as a function of distance from the body to other antennas a helix antenna has been designed. A helix

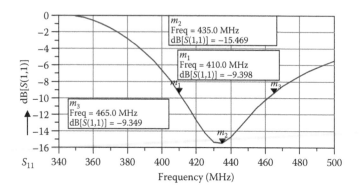

FIGURE 6.27 Measured S_{11} on the human body.

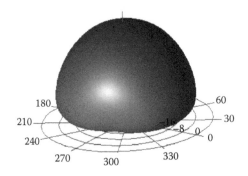

FIGURE 6.28 Compact antenna 3D radiation pattern.

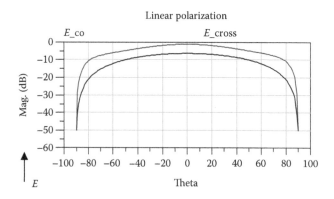

FIGURE 6.29 Antenna radiation pattern.

antenna with nine turns is shown in Figure 6.30. The backside of the circuit is copper under the microstrip matching stubs. However, in the helix antenna area there is no ground plane. The antenna has been designed to operate on the human body. A matching microstrip line network has been designed on a 0.8 mm thick RO4003 substrate. The helix antenna has VSWR better than 3:1 at the frequency range from

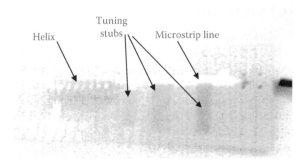

FIGURE 6.30   Helix antenna for medical applications.

440 MHz to 460 MHz. The antenna dimensions are $4 \times 4 \times 0.6$ cm. Figure 6.31 presents the measured S_{11} parameters on the human body. The computed E and H radiation planes of the helix antenna are shown in Figure 6.32. The helix antenna input impedance variation as a function of distance from the body is very sensitive. If the air spacing between the helix antenna and the human body is increased from 0 mm to 2 mm the antenna resonant frequency is shifted by 5%.

However, if the air spacing between the new dual polarized antenna and the human body is increased from 0 mm to 5 mm the antenna resonant frequency is shifted only by 5%.

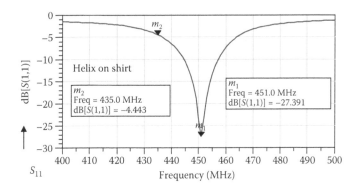

FIGURE 6.31   Measured S_{11} on the human body.

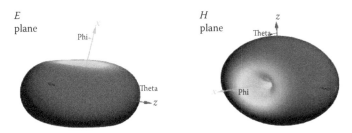

FIGURE 6.32 E and H plane radiation pattern of the helix antenna.

6.8 COMPACT WEARABLE RFID ANTENNAS

Radio frequency identification (RFID) is an electronic method of exchanging data over RF waves. There are three major components in the RFID system: transponder (tag), antenna, and a controller. The RFID tag, antenna, and controller may be assembled on the same board. Microstrip antennas have been widely presented in books and papers in the last decade [1–7]. However, compact wearable printed antennas are not widely used at 13.5 MHz RIFD systems. High-frequency tags work best at close range but are more effective at penetrating nonmetal objects, especially objects with high water content. A new class of wideband compact printed and microstrip antennas for RFID applications is presented in this chapter. RF transmission properties of human tissues have been investigated in several papers [8,9]. The effect of the human body on the antenna performance is investigated in this chapter. The proposed antennas may be used as wearable antennas on persons or animals and may be attached to cars, trucks, containers, and other various objects.

6.8.1 Dual Polarized 13.5-MHz Compact Printed Antenna

One of the most critical elements of any RFID system is the electrical performance of its antenna. The antenna is the main component for transferring energy from the transmitter to the passive RFID tags, receiving the transponder's replying signal and avoiding in-band interference from electrical noise and other nearby RFID components. Low-profile compact printed antennas are crucial in the development of RFID systems.

A new compact microstrip loaded dipole antenna has been designed at 13.5 MHz to provide horizontal polarization. The antenna consists of two layers. The first layer consists of a 0.8 mm FR4 dielectric substrate. The second layer consists of a 0.8 mm Kapton dielectric substrate. The substrate thickness determines the antenna bandwidth. A printed slot antenna provides a vertical polarization. The proposed antenna is dual polarized. The printed dipole and the slot antenna provide dual orthogonal polarizations. The dual polarized RFID antenna is shown in Figure 6.33. The antenna dimensions are

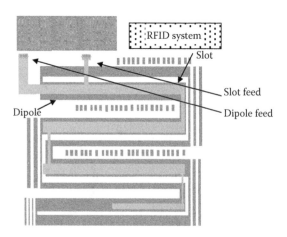

FIGURE 6.33   Printed compact dual polarized antenna, 64 × 64 × 1.6 mm.

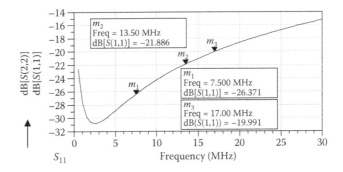

FIGURE 6.34 Computed S_{11} results.

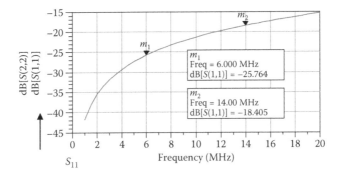

FIGURE 6.35 Measured S_{11} on the human body.

$6.4 \times 6.4 \times 0.16$ cm. The antenna may be attached to the customer's shirt in the stomach or back zone. The antenna has been analyzed by using Agilent ADS software.

The antenna S_{11} parameter is better than -21 dB at 13.5 MHz. The antenna gain is around -10 dBi. The antenna beam width is around $160°$. The computed S_{11} parameters are presented in Figure 6.34. There is a good agreement between measured and computed results. Figure 6.35 presents the antenna measured S_{11} parameters. The antenna cross-polarized field strength may be adjusted by varying the slot feed location. The computed radiation pattern is shown in Figure 6.36. The computed 3D radiation pattern of the antenna is shown in Figure 6.37.

6.8.2 VARYING THE ANTENNA FEED NETWORK

Several designs with different feed network have been developed. A compact antenna with different feed network is shown in Figure 6.38. The antenna dimensions are $8.4 \times 6.4 \times 0.16$ cm. Figure 6.39 presents the antenna computed S_{11} on the human body. There is a good agreement between measured and computed results. The computed radiation pattern is shown in Figure 6.40. Table 6.5 compares the electrical performance of a loop antenna with that of the compact dual polarized antenna.

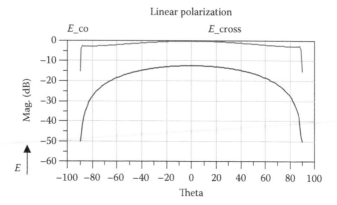

FIGURE 6.36 Antenna radiation pattern.

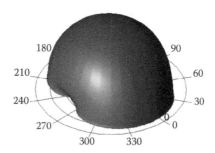

FIGURE 6.37 3D antenna radiation pattern.

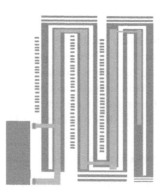

FIGURE 6.38 RFID printed antenna, 8.4 × 6.4 × 0.16 cm.

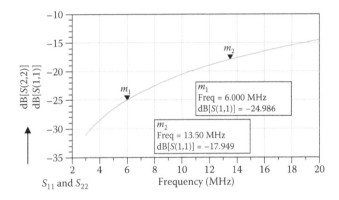

FIGURE 6.39 RFID antenna computed S_{11} and S_{22} results.

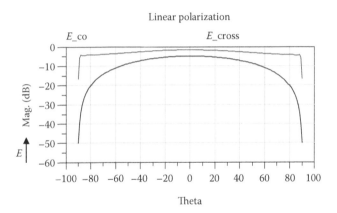

FIGURE 6.40 Compact antenna, $8.4 \times 6.4 \times 0.16$ cm, radiation pattern.

TABLE 6.5

Comparison of Loop Antenna and Microstrip Antenna Parameters

Antenna	Beam Width 3 dB (°)	Gain (dBi)	VSWR
Loop antenna	140	−25	2:1
Microstrip antenna	160	−10	1.2:1

6.8.3 RFID WEARABLE LOOP ANTENNAS

Several RFID loop antennas are presented in Ref. [16]. The disadvantages of loop antennas with a number of turns are low efficiency and narrow bandwidth. The real part of the loop antenna impedance approaches 0.5 Ω. The image part of the loop antenna impedance may be represented as high inductance. A matching network may be used to match the antenna to 50 Ω. The matching network consists of an RLC matching network. This matching network has a narrow bandwidth. The loop antenna efficiency is lower than 1%.

A square four-turn loop antenna has been designed at 13.5 MHz by using Agilent ADS software. The antenna is printed on an FR4 substrate. The antenna dimensions are 32 × 52.4 × 0.25 mm. The antenna layout is shown in Figure 6.41a, and the antenna photo is shown in Figure 6.41b. S_{11} results of the printed loop antenna are shown in Figure 6.42. The antenna S_{11} parameter is better than −9.5 dB without an external matching network. The computed radiation pattern is shown in Figure 6.43. The computed radiation pattern takes into account an infinite ground plane.

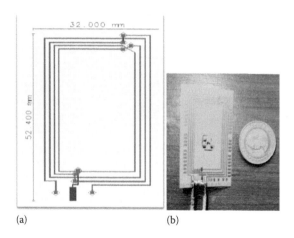

(a)   (b)

FIGURE 6.41 A square four turn loop antenna: (a) layout and (b) photo.

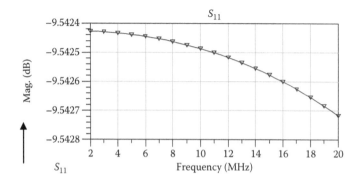

FIGURE 6.42 Loop antenna computed S_{11} results.

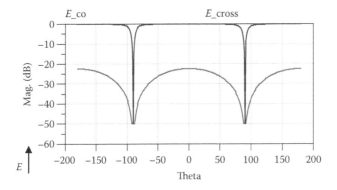

FIGURE 6.43 Loop antenna radiation patterns for an infinite ground plane.

The microstrip antenna input impedance variation as a function of distance from the body has been computed by employing ADS software. The analyzed structure is presented in Figure 6.17. Properties of human body tissues are listed in Table 6.2; see Ref. [8]. These properties were used in the antenna design. S_{11} parameters for different human body thicknesses have been computed. We may note that the differences in the results for body thickness of 15 mm to 100 mm are negligible. S_{11} parameters for different positions relative to the human body have been computed. If the air spacing between the antenna and the human body is increased from 0 mm to 10 mm the antenna S_{11} parameters may change by less than 1%. The VSWR is better than 1.5:1.

6.8.4 PROPOSED ANTENNA APPLICATIONS

An application of the proposed antenna is shown in Figure 6.44. The RFID antennas may be assembled in a belt and attached to the customer's stomach. The antennas may be employed as transmitting or as receiving antennas. The antennas may receive or transmit information to medical systems.

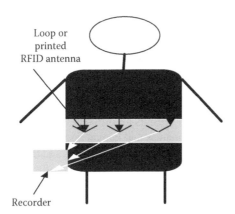

FIGURE 6.44 Wearable RFID antenna.

FIGURE 6.45   New microstrip antenna for RFID applications.

FIGURE 6.46   Loop antenna for RFID applications.

In RFID systems the distance between the transmitting and receiving antennas is less than $2D^2/\lambda$, where D is the largest dimension of the antenna. The receiving and transmitting antennas are magnetically coupled. In these applications we refer to the near-field and not to the far-field radiation pattern.

Figures 6.45 and 6.46 present compact printed antenna for RFID applications. The presented antennas may be assembled in a belt and attached to the patient's stomach or back.

6.9   CONCLUSIONS

This chapter presents wideband microstrip antennas with high efficiency for medical applications. The antenna dimensions may vary from $26 \times 6 \times 0.16$ cm to $5 \times 5 \times 0.05$ cm according to the medical system specification. The antenna bandwidth is around 10% for VSWR better than 2:1. The antenna beam width is around 100°. The antennas gain varies from 0 to 4 dBi. The antenna S_{11} results for different belt thicknesses, shirt thicknesses, and air spacing between the antennas and human body are presented in this chapter. If the air spacing between the new dual polarized antenna and the human body is increased from 0 mm to 5 mm the antenna resonant frequency is shifted by 5%. However, if the air spacing between the helix antenna and the human body is increased only from 0 mm to 2 mm the antenna resonant frequency is shifted by 5%. The effect of the antenna location on the human body should be considered in the antenna design process. The proposed antenna may be used in Medicare RF systems.

A wideband tunable microstrip antenna with high efficiency for medical applications has been presented in this chapter. The antenna dimensions may vary from $26 \times 6 \times 0.16$ cm to $5 \times 5 \times 0.05$ cm according to the medical system specification. The antenna's bandwidth is around 10% for VSWR better than 2:1. The antenna beam width is around 100°. The antenna's gain varies from 0 to 2 dBi. If the air spacing between the dual polarized antenna and the human body is increased from 0 mm to 5 mm the antenna resonant frequency is shifted by 5%. A varactor is employed to compensate variations in the antenna resonant frequency at different locations on the human body.

This chapter presents also wideband compact printed antennas, microstrip and loop antennas, for RFID applications. The antenna beam width is around 160°. The antenna gain is around –10 dBi. The proposed antennas may be used as wearable antennas on persons or animals. The proposed antennas may be attached to cars, trucks, and various other objects. If the air spacing between the antenna and the human body is increased from 0 mm to 10 mm the antenna S_{11} parameters may change by less than 1%. The antenna VSWR is better than 1.5:1 for all tested environments.

REFERENCES

1. James, J. R., Hall, P. S., and Wood, C. *Microstrip Antenna Theory and Design*, London: The Institution of Engineering and Technology, 1981.
2. Sabban, A., and Gupta, K. C. Characterization of Radiation Loss from Microstrip Discontinuities Using a Multiport Network Modeling Approach. *IEEE Transactions on Microwave Theory and Techniques*, 39(4): 705–712.
3. Sabban, A. A New Wideband Stacked Microstrip Antenna. *IEEE Antenna and Propagation Symposium*, Houston, TX, June 1983.
4. Sabban, A., and Navon, E. A MM-Waves Microstrip Antenna Array. In *IEEE Symposium*, Tel Aviv, Israel, March 1983.
5. Kastner, R., Heyman, E., Sabban, A. Spectral Domain Iterative Analysis of Single and Double-Layered Microstrip Antennas Using the Conjugate Gradient Algorithm. *IEEE Transactions on Antennas and Propagation*, 36(9): 1204–1212, 1988.
6. Sabban, A. Wideband Microstrip Antenna Arrays. *IEEE Antenna and Propagation Symposium MELCOM*, Tel Aviv, Israel, June 1981.
7. Sabban, A. Microstrip Antenna Arrays. In N. Nasimuddin (Ed.), *Microstrip Antennas*, pp. 361–384, 2011. InTech, http://www.intechopen.com/articles/show/titl/microstrip-antenna-arrays.
8. Chirwa, L. C., Hammond, P. A., Roy, S., and Cumming, D. R. S. Electromagnetic Radiation from Ingested Sources in the Human Intestine between 150 MHz and 1.2 GHz. *IEEE Transactions on Biomedical Engineering*, 50(4): 484–492, 2003.
9. Werber, D., Schwentner, A., and Biebl, E. M. Investigation of RF Transmission Properties of Human Tissues. *Advances in Radio Science*, 4: 357–360, 2006.
10. Gupta, B., Sankaralingam S., and Dhar, S. Development of Wearable and Implantable Antennas in the Last Decade. In *Microwave Symposium (MMS)*, Mediterranean 2010, pp. 251–267.
11. Thalmann, T., Popovic, Z., Notaros, B. M., and Mosig, J. R. Investigation and Design of a Multi-Band Wearable Antenna. In *3rd European Conference on Antennas and Propagation, EuCAP2009*, pp. 462–465.
12. Salonen, P., Rahmat-Samii, Y., and Kivikoski, M. Wearable Antennas in the Vicinity of Human Body. *IEEE Antennas and Propagation Society International Symposium*, 1: 467–470, 2004.

7 Wearable Tunable Printed Antennas for Medical Applications

7.1 INTRODUCTION

Communication and biomedical industry is in continuous growth in the last decade. Low-profile compact tunable antennas are crucial in the development of wearable human biomedical systems. Tunable antennas consist of a radiating element and of a voltage-controlled diode, varactor. Varactor diodes are semiconductor devices that are voltage-controlled variable capacitance. The radiating element may be a microstrip patch antenna, dipole, or loop antenna. The antenna resonant frequency may be tuned by using a varactor to compensate variations in the antenna resonant frequency.

Microstrip antenna resonant frequency is altered due to environment condition, different antenna locations, and different system mode of operation. Printed tunable antennas are not widely presented. However, microstip antennas are widely presented in books and papers in the last decade [1–7]. A new class of wideband tunable wearable microstrip antennas for medical applications is presented in this chapter.

7.2 VARACTOR THEORY

Varactor diodes are semiconductor devices that are used in many microwave systems where a voltage-controlled variable capacitance is required.

Varactor p–n junction diodes exhibit a variable capacitance effect and may be used as a voltage-controlled variable capacitance. However, special p–n diodes are optimized and fabricated to give the required capacitance values. Varactor diodes normally enable much higher ranges of capacitance change to be achieved as a result of the p–n diode optimized design. Varactor diodes are widely used in radio frequency (RF) devices. The circuit capacitance is varied by applying a controlled voltage. Varactor diodes are used in voltage-controlled oscillators (VCOs). Varactor diodes are used also in tunable filters and antennas.

7.2.1 VARACTOR DIODE BASICS

The varactor diode consists of a standard PN junction as shown in Figure 7.1. The diode is operated under reverse bias conditions, which gives rise to three regions and there is no conduction. The left and right ends of the diode are p and n regions, where current can be conducted. However around the junction is the depletion region

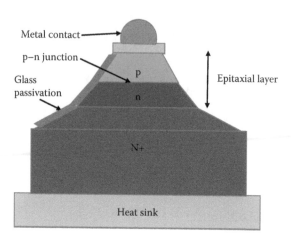

FIGURE 7.1   Internal structure of a varactor.

where no current carriers are available. As a result, current can be carried in the p
and n regions, but the depletion region is an insulator. This is similar to a capacitor
structure. It has conductive plates separated by an insulating dielectric. The capacitance of a capacitor depends on the plate area, the dielectric constant of the insulator
between the plates, and the distance between the two plates. In the case of the varactor diode, it is possible to increase and decrease the width of the depletion region by
changing the level of the reverse bias. This has the effect of changing the distance
between the plates of the capacitor. However, to be able to use varactor diodes to
their best advantage it is necessary to understand features of varactor diodes including the capacitance ratio, Q, gamma, reverse voltage, and the like.

Varactors provide an electrically controllable capacitance, which can be used in
tuned circuits. It is small and inexpensive. Its disadvantages compared to a manually
controlled variable capacitor are a lower Q, nonlinearity, lower voltage rating, and a
more limited range. A equivalent circuit of varactors is presented in Figure 7.2.

Any p–n junction has a junction capacitance that is a function of the voltage
across the junction. The electric field in the depletion layer that is set up by the ion-
ized donors and acceptors is responsible for the voltage difference that balances the
applied voltage. A higher reverse bias widens the depletion layer, uncovering more

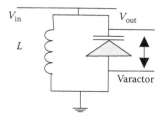

FIGURE 7.2   Varactor example.

fixed charge and raising the junction potential. The capacitance of the junction is $C = Q(V)/V$, and the incremental capacitance is $c = \Delta Q(V)/\Delta V$. The capacitance to be used in the formula for the resonant frequency is the incremental capacitance, where it is assumed that the incremental voltage ΔV is small compared to V. Finite voltages give rise to nonlinearities. The capacitance decreases as the reverse bias increases, according to the relation $C = C_0/(1 + V/V_0)^n$, where C_0 and V_0 are constants. The diode forward voltage is approximately V_0. The exponent n depends on how the doping density of the semiconductors relates to distance away from the junction. For a graded junction (linear variation), $n = 0.33$. For an abrupt junction (constant doping density), $n = 0.5$. If the density jumps abruptly at the junction, then decreases (called hyperabrupt), n can be made as high as $n = 2$. The varactor capacitance is given in Equation 7.1. The circuit frequency f_r may be calculated by using Equation 7.2.

$$C = \frac{A\varepsilon}{d} \tag{7.1}$$

where
   C = capacitance
   A = plate area
   d = diode thickness

$$f_r = \frac{1}{2\pi\sqrt{LC}} \tag{7.2}$$

7.2.2   Types of Varactors

An *abrupt varactor* is a varactor in which the changeover p–n junction is abrupt. In a *hyperabrupt varactor,* the change is very abrupt. Varactors are used in oscillators to sweep for different frequencies.

   In *gallium-arsenide varactor diodes* the semiconductor material used is gallium arsenide. These types are used for frequencies from 18 GHz up to and beyond 600 GHz.

7.3   DUALLY POLARIZED TUNABLE PRINTED ANTENNA

A compact tunable microstrip dipole antenna has been designed to provide horizontal polarization. The antenna consists of two layers. The first layer consists of a 0.8 mm thick RO3035 dielectric substrate. The second layer consists of a 0.8 mm thick RT/Duroid® 5880 dielectric substrate. The substrate thickness affects the antenna band width. The printed slot antenna provides a vertical polarization. The printed dipole and the slot antenna provide dual orthogonal polarizations. The dimensions of the dual polarized antenna are 26 × 6 × 0.16 cm. Tunable compact folded dual polarized antennas have also been designed. The dimensions of the compact antennas are

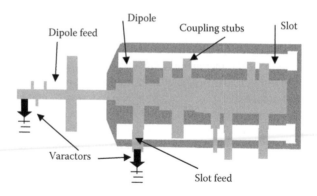

FIGURE 7.3   Dual polarized tunable antenna, 26 × 6 × 0.16 cm.

5 × 5 × 0.05 cm. Varactors are connected to the antenna feed lines as shown in Figure 7.3. The voltage-controlled varactors are used to control the antenna resonant frequency. The varactor bias voltage may be varied automatically to set the antenna resonant frequency at different locations on the human body. The antenna may be used as a wearable antenna on the human body. It may be attached to the patient's shirt in the stomach or back zone. The antenna has been analyzed by using Agilent ADS software. There is a good agreement between measured and computed results. The antenna bandwidth is around 10% for voltage standing wave ratio (VSWR) better than 2:1. The antenna beam width is around 100°. The antenna gain is around 2 dBi.

Figure 7.4 presents the antenna measured S_{11} parameters without a varactor. Figure 7.5 presents the antenna S_{11} parameters as a function of different varactor capacitances. Figure 7.6 presents the tunable antenna resonant frequency as a function of the varactor capacitance. The antenna resonant frequency varies around 5% for capacitances up to 2.5 pF. The antenna beam width is 100°. The antenna cross-polarized field strength may be adjusted by varying the slot feed location.

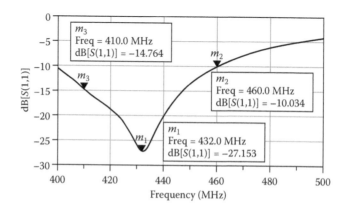

FIGURE 7.4   Measured S_{11} on human body.

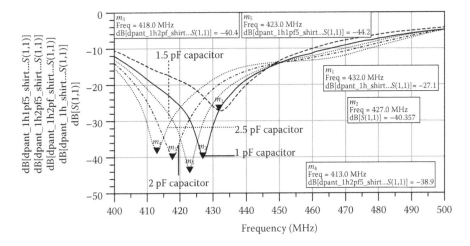

FIGURE 7.5 The tunable S_{11} parameter as function of varactor capacitance.

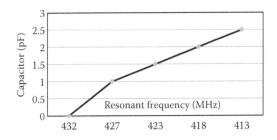

FIGURE 7.6 Resonant frequency as a function of varactor capacitance.

7.4 WEARABLE TUNABLE ANTENNAS

As presented in Chapter 6, the antenna input impedance varies as a function of distance from the body. Wearable antennas' electrical performance was computed by using ADS software [8,9]. Wearable antennas are presented in books and papers in the last decade [10–16]. However, wearable tunable antennas are not widely presented. In this section, several wearable tunable antennas are presented. The analyzed structure is presented in Figure 6.14. Properties of human body tissues are listed in Table 6.2 (see Ref. [8]). Figure 7.7 presents S_{11} results for different belt and shirt thickness, and air spacing between the antennas and the human body. When the antenna resonant frequency is shifted, the voltage across the varactor is varied to tune the antenna resonant frequency.

If the air spacing between the sensors and the human body is increased from 0 to 5 mm, the antenna resonant frequency is shifted by 5%. A voltage-controlled varactor is used to tune the antenna resonant frequency due to different antenna locations on a human body. Figure 7.8 presents several compact tunable antennas for medical applications. A voltage-controlled varactor may be used also to tune the loop antenna resonant frequency at different antenna locations on the body.

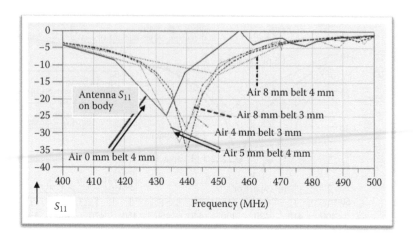

FIGURE 7.7 S_{11} of the antenna for different spacing relative to the human body.

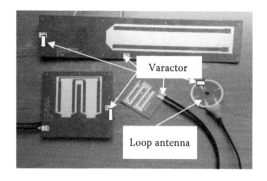

FIGURE 7.8 Tunable antennas for medical applications.

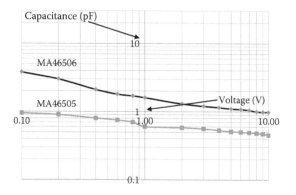

FIGURE 7.9 Varactor capacitance as a function of bias voltage.

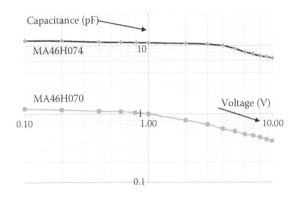

FIGURE 7.10   *C–V* curves of varactors MA46H070 to MA46H074.

7.5   TUNABLE ANTENNA VARACTORS

Tuning varactors are variable capacitors designed to provide electronic tuning of microwave components. Varactors are manufactured on silicon and gallium arsenide substrates. Gallium arsenide varactors offer higher Q and may be used at higher frequencies than silicon varactors. Hyperabrupt varactors provide nearly linear variation of frequency with applied control voltage. However, abrupt varactors provide inverse fourth root frequency dependence. MACOM offers several gallium arsenide hyperabrupt varactors such as the MA46 series. Figure 7.9 presents the *C–V* curves of varactors MA46505 to MA46506. Figure 7.10 presents the *C–V* curves of varactors MA46H070 to MA46H074.

7.6   MEASUREMENTS OF TUNABLE ANTENNAS

Figure 7.11 presents a compact tunable antenna with a varactor. A varactor was connected to the antenna feed line. The varactor bias voltage was varied from 0 V to 9 V. Figure 7.12 presents measured S_{11} as a function of varactor bias voltage. The

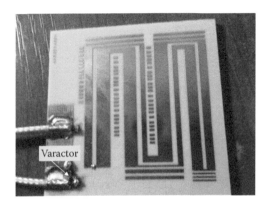

FIGURE 7.11   Tunable antenna with a varactor.

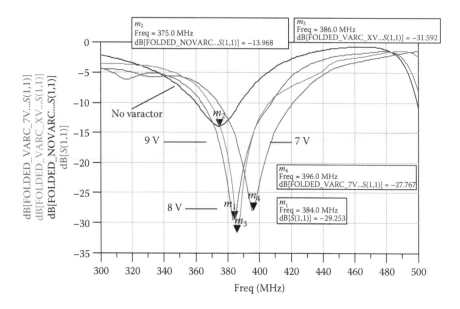

FIGURE 7.12 Measured S_{11} as function of varactor bias voltage.

antenna resonant frequency was shifted by 5% for bias voltage between 7 V and 9 V. We may conclude that varactors may be used to compensate variations in the antenna resonant frequency at different locations on the human body.

7.7 FOLDED DUAL POLARIZED TUNABLE ANTENNA

The dimensions of the folded dual polarized antenna presented in Figure 7.13 are $7 \times 5 \times 0.16$ cm. The length and width of the coupling stubs in Figure 7.13 are 12 mm by 9 mm. Small tuning bars are located along the feed line to tune the

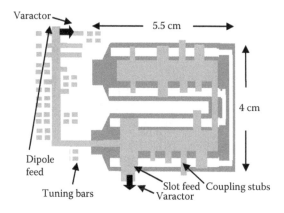

FIGURE 7.13 Tunable folded dual polarized antenna.

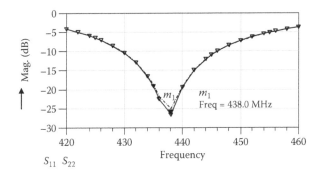

FIGURE 7.14 Folded antenna computed S_{11} and S_{22} results.

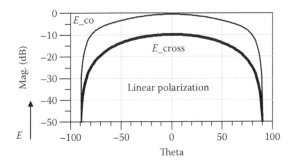

FIGURE 7.15 Folded antenna radiation pattern.

antenna to the desired resonant frequency. Figure 7.14 presents the antenna computed S_{11} and S_{22} parameters. The computed radiation pattern of the folded dipole is shown in Figure 7.15.

7.8 MEDICAL APPLICATIONS FOR TUNABLE ANTENNAS

Three to four tunable folded dipoles or tunable loop antennas may be assembled in a belt and attached to the patient stomach or back, as shown in Figure 7.16. The bias voltage to the varactors is supplied by a recorder battery. The RF and DC cables from each antenna are connected to a recorder. The received signal is routed to a switching matrix. The signal with the highest level is selected during the medical test. The varactor bias voltage may be varied to tune the antenna resonant frequency. The antennas receive a signal that is transmitted from various positions in the human body. Tunable antenna may be attached to the patient's back in order to improve the level of the received signal from different locations in the human body. In several applications, the distance separating the transmitting and receiving antennas is less than the far-field distance, $2D^2/\lambda$, where D is the largest dimension of the source of the radiation, and λ is the wavelength. In these applications, the amplitude of the

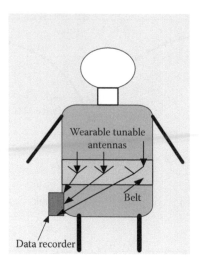

FIGURE 7.16   Tunable wearable antenna.

electromagnetic field close to the antenna falls off rapidly with the distance from the antenna. The electromagnetic fields do not radiate energy to infinite distances, but instead their energies remain trapped in the antenna near zone. The near-fields transfer energy only to close distances from the receivers. In these applications, we have to refer to the near-field and not to the far-field radiation. The receiving and transmitting antennas are magnetically coupled. Change in current flow through one wire induces a voltage across the ends of the other wire through electromagnetic induction. The proposed tunable wearable antennas may be placed on the patient body as

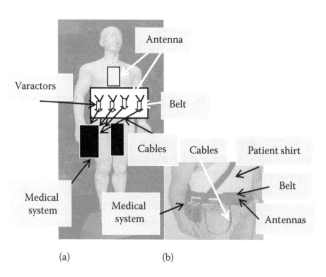

FIGURE 7.17   (a) Medical system with printed wearable antennas. (b) Patient with printed wearable antenna.

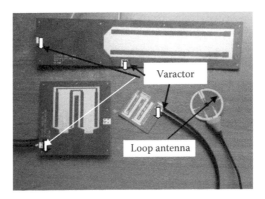

FIGURE 7.18 Tunable antennas for medical applications.

shown in Figure 7.17a. The patient in Figure 7.17b is wearing a wearable antenna. The antenna belt is attached to the patient's front or back body. Figure 7.18 presents several compact tunable antennas for medical applications. A voltage-controlled varactor may be used also to tune the wearable antenna resonant frequency at different antenna locations on the body.

7.9 CONCLUSIONS

This chapter described wideband tunable microstrip antennas with high efficiency for medical applications. The antenna dimensions may vary from 26 cm by 6 cm by 0.16 cm to 5 cm by 5 cm by 0.05 cm according to the medical system specification. The antenna's bandwidth is around 10% for VSWR better than 2:1. The antenna beam width is around 100°. The antenna's gain varies from 0 to 2 dBi. If the air spacing between the dual polarized antenna and the human body is increased from 0 mm to 5 mm the antenna resonant frequency is shifted by 5%. A varactor is employed to compensate variations in the antenna resonant frequency at different locations on the human body.

REFERENCES

1. James, J. R., Hall, P. S., and Wood, C. *Microstrip Antenna Theory and Design*. London: The Institution of Engineering and Technology, 1981.
2. Sabban, A., and Gupta, K. C. Characterization of Radiation Loss from Microstrip Discontinuities Using a Multiport Network Modeling Approach. *IEEE Transactions on Microwave Theory and Techniques*, 39(4): 705–712, 1991.
3. Sabban, A. A New Wideband Stacked Microstrip Antenna. In *IEEE Antenna and Propagation Symposium*, Houston, TX, June 1983.
4. Sabban, A., and Navon, E. A MM-Waves Microstrip Antenna Array. In *IEEE Symposium*, Tel Aviv, Israel, March 1983.
5. Kastner, R., Heyman, E., and Sabban, A. Spectral Domain Iterative Analysis of Single and Double-Layered Microstrip Antennas Using the Conjugate Gradient Algorithm. *IEEE Transactions on Antennas and Propagation*, 36(9):1204–1212, 1988.
6. Sabban, A. Wideband Microstrip Antenna Arrays. In *IEEE Antenna and Propagation Symposium MELCOM*, Tel Aviv, Israel, June 1981.

7. Sabban, A. Microstrip Antenna Arrays. In N. Nasimuddin (Ed.), *Microstrip Antennas.*, pp. 361–384, 2011. InTech, http://www.intechopen.com/articles/show/title/microstrip -antenna-arrays.
8. Chirwa, L. C., Hammond, P. A., Roy, S., and Cumming, D. R. S. Electromagnetic Radiation from Ingested Sources in the Human Intestine between 150 MHz and 1.2 GHz. *IEEE Transaction on Biomedical Engineering*, 50(4):484–492, 2003.
9. Werber, D., Schwentner, A., and Biebl, E. M. Investigation of RF Transmission Properties of Human Tissues. *Advances in Radio Science*, 4:357–360, 2006.
10. Gupta, B., Sankaralingam S., and Dhar, S. Development of Wearable and Implantable Antennas in the Last Decade. In *Microwave Mediterranean Symposium (MMS)*, August 2010, Cyprus, pp. 251–267.
11. Thalmann, T., Popovic, Z., Notaros, B. M., and Mosig, J. R. Investigation and Design of a Multi-Band Wearable Antenna. In *3rd European Conference on Antennas and Propagation, EuCAP*, March 2009, Berlin, Germany, pp. 462–465.
12. Salonen, P., Rahmat-Samii, Y., and Kivikoski, M. Wearable Antennas in the Vicinity of Human Body. In *IEEE Antennas and Propagation Society International Symposium*, Monterey, California, 2004, vol. 1, pp. 467–470.
13. Kellomaki, T., Heikkinen, J., and Kivikoski, M. *Antennas and Propagation, EuCAP 2006*, November 2006, Nice, France, pp. 1–6.
14. Sabban, A. Wideband Printed Antennas for Medical Applications. In APMC 2009 Conference, Singapore, December 2009.
15. Lee, L. Antenna Circuit Design for RFID Applications. Microchip Technology Inc., Microchip AN 710c.
16. ADS software, Agilent. http://www.home.agilent.com/agilent/product.jspx?cc=IL&lc =eng&ckey=1297113&nid=-34346.0.00&id=1297113.

8 New Wideband Wearable Meta-Material Antennas for Communication Applications

8.1 INTRODUCTION

Microstrip antennas are widely used in communication systems. They have several advantages such as low profile, flexibility, light weight, small volume, and low production cost. Compact printed antennas are discussed in journals and books [1–4]. However, small printed antennas suffer from low efficiency. Meta material technology is used to design small printed antennas with high efficiency. Printed wearable antennas are described in Ref. [5]. Artificial media with negative dielectric permittivity are described in Ref. [6]. Periodic split-ring resonator (SRR) and metallic post structures may be used to design materials with dielectric constant and permeability less than 1 as described in Refs. [6–14]. This chapter discusses use of meta-material technology to develop small antennas with high efficiency. Radio frequency transmission properties of human tissues have been investigated in several papers such as Refs. [15,16]. Several wearable antennas have been presented in papers in the last few years [17–24]. New wearable printed meta-material antennas with high efficiency are discussed in this chapter. The bandwidth of the meta-material antenna with SRR and metallic strips is around 50% for voltage standing wave ratio (VSWR) better than 2.3:1. Computed and measured results of meta-material antennas on the human body are discussed in this chapter.

8.2 NEW ANTENNAS WITH SRRs

A microstrip dipole antenna with SRRs is shown in Figure 8.1. A microstrip-loaded dipole antenna with SRR provides horizontal polarization. The slot antenna provides vertical polarization. The resonant frequency of the antenna with SRR is 400 MHz. The resonant frequency of the antenna without SRR is 10% higher. The antennas shown in Figure 8.1 consist of two layers. The dipole feed network is printed on the first layer. The radiating dipole with SRR is printed on the second layer. The thickness of each layer is 0.8 mm. The dipole and the slot antenna create dual polarized antenna. The computed S_{11} parameters are presented in Figure 8.2.

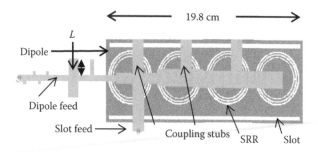

FIGURE 8.1   Printed antenna with SRRs.

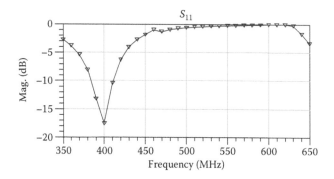

FIGURE 8.2   Antenna with SRRs, computed S_{11}.

The length of the dual polarized antenna with SRR shown in Figure 8.1 is 19.8 cm. The length of the dual polarized antenna without SRR shown in Figure 8.3 is 21 cm. The ring width is 1.4 mm; the spacing between the rings is 1.4 mm. The antennas were analyzed by using Agilent ADS software. The matching stub locations and dimensions have been optimized to get the best VSWR results. The length of stub L in Figures 8.1 and 8.3 is 10 mm. The locations and numbers of the coupling stubs may vary the antenna axial ratio from 0 dB to 30 dB. The number of coupling stubs may be minimized. The number of coupling stubs in Figure 8.1 is three. The antenna axial ratio value may be adjusted also by varying the slot feed

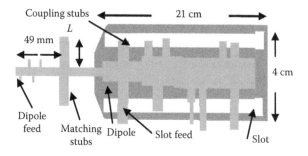

FIGURE 8.3   Dual polarized microstrip antenna.

location. The dimensions of the antenna shown in Figure 8.3 are presented in Ref. [5]. The bandwidth of the antenna shown in Figure 8.3 is around 10% for VSWR better than 2:1. The antenna beam width is 100°. The antenna gain is around 2 dBi. The computed S_{11} parameters are presented in Figure 8.4. Figure 8.5 presents the measured S_{11} parameters of the antenna. There is a good agreement between measured and computed results. The antenna presented in Figure 8.1 has been modified as shown in Figure 8.6. The location and the dimension of the coupling stubs have been modified to get two resonant frequencies. The first resonant frequency is 370 MHz and is lower by 20% than the resonant frequency of the antenna without the SRR.

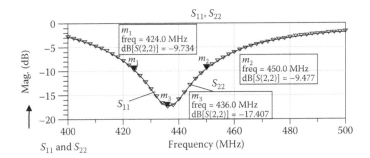

FIGURE 8.4 Computed S_1 and S_{22} results, antenna without SRR.

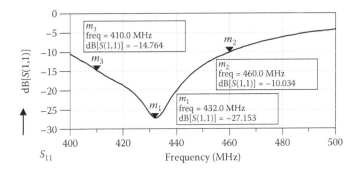

FIGURE 8.5 Measured S_{11} of the antenna without SRR.

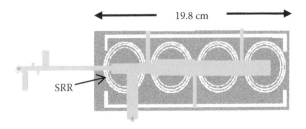

FIGURE 8.6 Antenna with SRR with two resonant frequencies.

The S_{11} for an antenna with two resonant frequencies is shown in Figure 8.7. Metallic strips have been added to the antenna with SRR as shown in Figure 8.8, and the S_{11} for this antenna is shown in Figure 8.9. The bandwidth is around 50% for VSWR better than 3:1. The computed radiation pattern is shown in Figure 8.10. The 3D computed radiation pattern is shown in Figure 8.11. The directivity and gain of the antenna with SRR is around 5 dBi (Figure 8.12). The directivity of the antenna without SRR is around 2 dBi. The length of the antennas with SRR is smaller by 5%

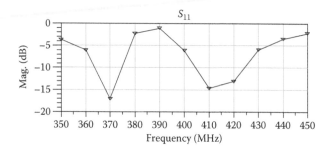

FIGURE 8.7 S_{11} for antenna with two resonant frequencies.

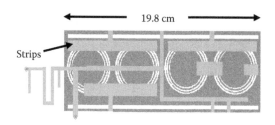

FIGURE 8.8 Antenna with SRR and metallic strips.

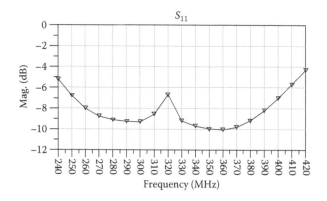

FIGURE 8.9 S_{11} for antenna with SRR and metallic strips.

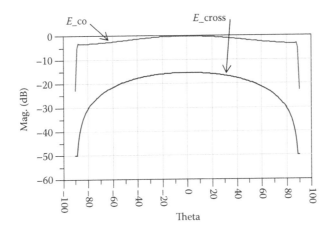

FIGURE 8.10   Radiation pattern for antenna with SRR.

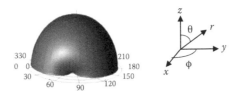

FIGURE 8.11   3D radiation pattern for antenna with SRR.

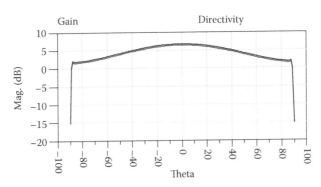

FIGURE 8.12   Directivity of the antenna with SRR.

than that of antennas without SRR. Moreover, the resonant frequency of the antennas with SRR is lower by 5%–10%.

The feed network of the antenna presented in Figure 8.8 has been optimized to yield VSWR better than 2:1 in the frequency range of 250 MHz to 440 MHz. Optimization of the number of the coupling stubs and the distance between the coupling stubs may be used to tune the antenna resonant frequency. An optimized antenna with two coupling stubs has two resonant frequencies. The first resonant frequency is 370 MHz and the second is 420 MHz. An antenna with SRR with two coupling stubs is presented in Figure 8.13.

The computed S_{11} parameter of the antenna with two coupling stubs is presented in Figure 8.14. The 3D radiation pattern for an antenna with SRR and two coupling stubs is shown in Figure 8.15.

The antenna with metallic strips has been optimized to yield wider bandwidth as shown in Figure 8.16. The computed S_{11} parameter of the modified antenna with metallic strips is presented in Figure 8.17. The antenna bandwidth is around 50% for VSWR better than 2.3:1.

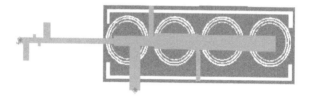

FIGURE 8.13   Antenna with SRR with two coupling stubs.

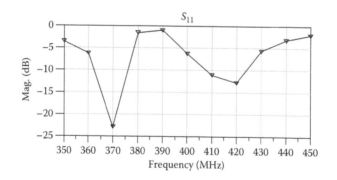

FIGURE 8.14 S_{11} for antenna with SRR with two coupling stubs.

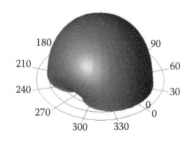

FIGURE 8.15   3D radiation pattern for antenna with SRR and two coupling stubs.

FIGURE 8.16   Wideband antenna with SRR and metallic strips.

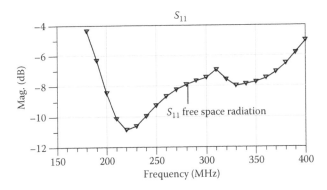

FIGURE 8.17 S_{11} for antenna with SRR and metallic strips.

8.3   FOLDED DIPOLE META-MATERIAL ANTENNA WITH SRRs

The length of the antenna shown in Figure 8.3 may be reduced from 21 cm to 7 cm by folding the printed dipole as shown in Figure 8.18. Tuning bars are located along the feed line to tune the antenna to the desired frequency. The antenna bandwidth is around 10% for VSWR better than 2:1 as shown in Figure 8.19. The antenna beam width is around 100°. The antenna gain is around 2 dBi. The size of the antenna with SRR shown in Figure 8.6 may be reduced by folding the printed dipole as shown in Figure 8.20. The dimensions of the folded dual polarized antenna with

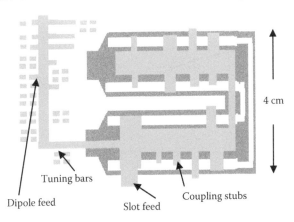

FIGURE 8.18   Folded dipole antenna, 7 × 5 × 0.16 cm.

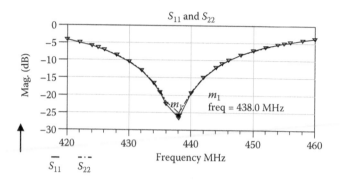

FIGURE 8.19   Folded antenna computed S_{11} and S_{22} results.

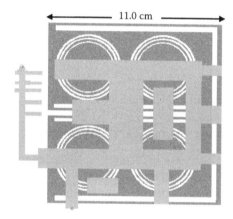

FIGURE 8.20   Folded dual polarized antenna with SRR.

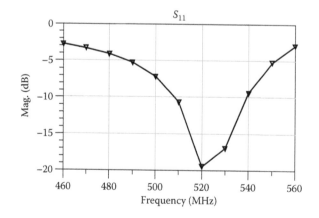

FIGURE 8.21   Folded antenna with SRR, computed S_{11}.

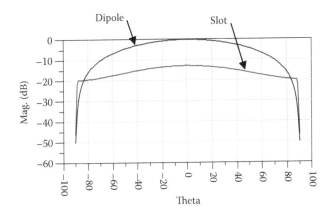

FIGURE 8.22 Radiation pattern of the folded antenna with SRR.

SRR presented in Figure 8.20 are 11 × 11 × 0.16 cm. Figure 8.21 presents the computed S_{11} antenna parameters. The antenna bandwidth is 10% for VSWR better than 2:1. The computed radiation pattern of the folded antenna with SRR is shown in Figure 8.22.

8.4 STACKED PATCH ANTENNA LOADED WITH SRRs

At first a microstrip stacked patch antenna [1–3] has been designed. The second step was to design the same antenna with SRR. The antenna consists of two layers. The first layer consists of a 1.6 mm thick FR4 dielectric substrate with dielectric constant of 4. The second layer consists of a 1.6 mm thick RT/Duroid® 5880 dielectric substrate with dielectric constant of 2.2. The dimensions of the microstrip stacked patch antenna shown in Figure 8.23 are 33 × 20 × 3.2 mm. The antenna has been analyzed by using Agilent ADS software. The antenna bandwidth is around 5% for VSWR better than 2.5:1. The antenna beam width is around 72°. The antenna gain is around 7 dBi. The computed S_{11} parameters are presented in Figure 8.24. The radiation

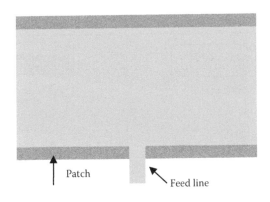

FIGURE 8.23 A microstrip stacked patch antenna.

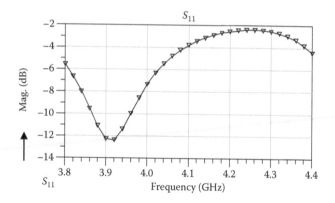

FIGURE 8.24 Computed S_{11} of the microstrip stacked patch.

pattern of the microstrip stacked patch is shown in Figure 8.25. The antenna with SRR is shown in Figure 8.26. This antenna has the same structure as the antenna shown in Figure 8.23. The ring width is 0.2 mm and the spacing between the rings is 0.25 mm. Twenty-eight SRR are placed on the radiating element. There is a good agreement between measured and computed results. The measured S_{11} parameters of the antenna with SRR are presented in Figure 8.27. The antenna bandwidth is around 12% for VSWR better than 2.5:1. By adding an air space of 4 mm between the antenna layers the VSWR was improved to 2:1. The antenna gain is around 9 to 10 dBi. The antenna efficiency is around 95%. The antenna computed radiation

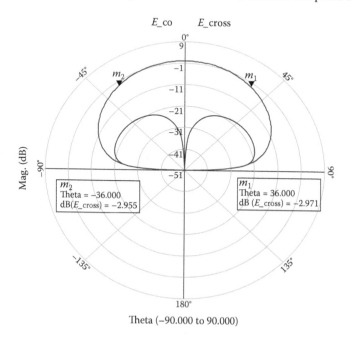

FIGURE 8.25 Radiation pattern of the microstrip stacked patch.

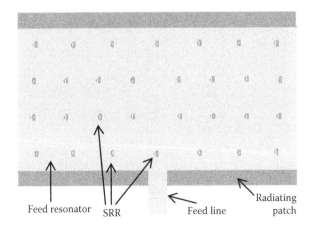

Feed resonator   SRR   Feed line   Radiating patch

FIGURE 8.26   Printed antenna with SRRs.

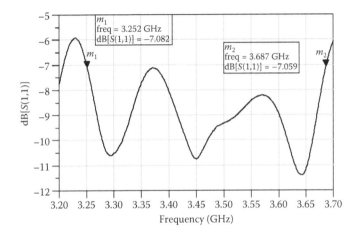

FIGURE 8.27   Patch with SRRs, measured S_{11}.

pattern is shown in Figure 8.28. The patch antenna with SRR performs as a loaded patch antenna. The effective area of a patch antenna with SRR is higher than the effective area of a patch antenna without SRR. The resonant frequency of a patch antenna with SRR is lower by 10% than the resonant frequency of a patch antenna without SRR. The antenna beam width is around 70°. The gain and directivity of the stacked patch antenna with SRR is higher by 2 dB to 3 dB than the patch antenna without SRR.

8.5   PATCH ANTENNA LOADED WITH SRRs

A patch antenna with SRRs has been designed. The antenna is printed on a 1.6 mm thick RT/Duroid 5880 dielectric substrate with dielectric constant of 2.2. The dimensions of the microstrip patch antenna shown in Figure 8.29 are 36 × 20 × 1.6 mm.

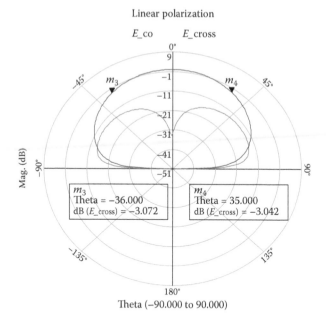

FIGURE 8.28   Radiation pattern for patch with SRR.

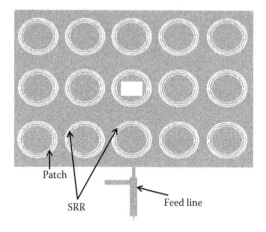

FIGURE 8.29   Patch antenna with SRRs.

The antenna bandwidth is around 5% for S_{11} lower than −9.5 dB. However, the antenna bandwidth is around 10% for VSWR better than 3:1. The antenna beam width is around 72°. The antenna gain is around 7.8 dBi. The directivity of the antenna is 8. The antenna gain is 6.03. The antenna efficiency is 77.25%. The computed S_{11} parameters are presented in Figure 8.30. The gain and directivity of the patch antenna with SRR is higher by 2.5 dB than the patch antenna without SRR.

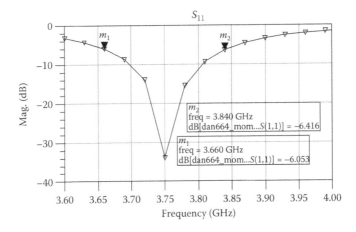

FIGURE 8.30   Patch with SRRs, computed S_{11}.

8.6   META-MATERIAL ANTENNA CHARACTERISTICS IN THE VICINITY OF THE HUMAN BODY

The antenna's input impedance variation as a function of distance from the human body had been computed using the structure presented in Figure 8.31. The electrical properties of human body tissues are listed in Table 8.1 (see Ref. [15]). The antenna location on the human body may be taken into account by calculating the S_{11} for different dielectric constants of the body. The variation of the dielectric constant of the body from 43 at the stomach to 63 at the colon zone shifts the antenna resonant frequency by 2%. The antenna was placed inside a belt with thickness between 1 and 4 mm with dielectric constant from 2 to 4. Figure 8.32 presents S_{11} results of the antenna with SRR on the human body shown in Figure 8.13. The antenna resonant frequency is shifted by 3%. Figure 8.33 presents S_{11} results of the antenna with SRR and metallic strips, shown in Figure 8.16, on the human body. The antenna resonant frequency is shifted by 1%. The air spacing between the belt and the patient shirt is varied from 0 mm to 8 mm. The dielectric constant of the patient shirt was varied from 2 to 4.

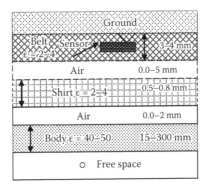

FIGURE 8.31   Wearable antenna environment.

TABLE 8.1

Dielectric Constant and Conductivity of Human Body Tissues

Tissue	Property	434 MHz	800 MHz	1000 MHz
Prostate	σ	0.75	0.90	1.02
	ε	50.53	47.4	46.65
Stomach	σ	0.67	0.79	0.97
	ε	42.9	40.40	39.06
Colon, Heart	σ	0.98	1.15	1.28
	ε	63.6	60.74	59.96
Kidney	σ	0.88	0.88	0.88
	ε	117.43	117.43	117.43
Nerve	σ	0.49	0.58	0.63
	ε	35.71	33.68	33.15
Lung	σ	0.27	0.27	0.27
	ε	38.4	38.4	38.4
Muscle	σ	0.98	1.15	1.28
	ε	63.6	60.74	59.96

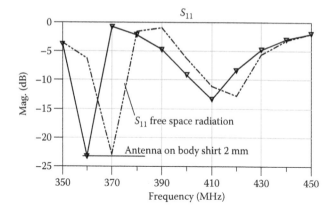

FIGURE 8.32 S_{11} of the antenna with SRR on the human body.

Figure 8.34 presents S_{11} results (of the antenna shown in Figure 8.3) for different air spacing between the antennas and human body, belt thicknesses, and shirt thicknesses. Results presented in Figure 8.33 indicate that the antenna has VSWR better than 2.5:1 for air spacing up to 8 mm between the antennas and the body. Figure 8.35 presents S_{11} results for different positions relative to the human body of the folded antenna shown in Figure 8.6. An explanation of Figure 8.35 is given in Table 8.2. If the air spacing between the antennas and the human body is increased from 0 mm to 5 mm the antenna resonant frequency is shifted by 5%. Tunable wearable antenna may be used to control the antenna resonant frequency at different positions on the human body (see Ref. [25]).

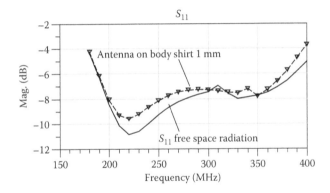

FIGURE 8.33 Antenna with SRR S_{11} results on the human body.

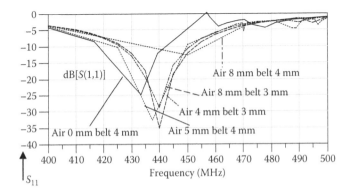

FIGURE 8.34 S_{11} results of the antenna shown in Figure 8.3 on the human body.

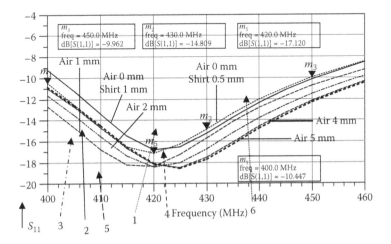

FIGURE 8.35 S_{11} results for different locations relative to the human body for the antenna shown in Figure 8.6.

TABLE 8.2

Explanation of Figure 8.35

Picture	Line Type	Sensor Position
1	Dot	Shirt thickness 0.5 mm
2	Line ____	Shirt thickness 1 mm
3	Dash dot _._.	Air spacing 2 mm
4	Dash	Air spacing 4 mm
5	Long dash _ _	Air spacing 1 mm
6	Big dots ••••••	Air spacing 5 mm

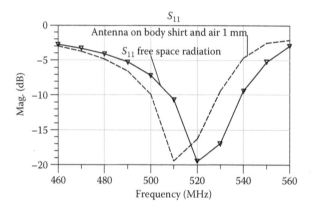

FIGURE 8.36 Folded antenna with SRR, S_{11} results on the body.

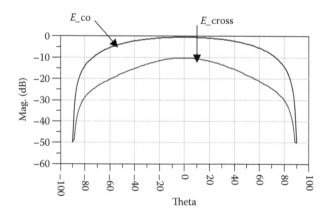

FIGURE 8.37 Radiation pattern of the folded antenna with SRR on the human body.

Figure 8.36 presents S_{11} results of the folded antenna with SRR, shown in Figure 8.8, on the human body. The antenna resonant frequency is shifted by 2%. The radiation pattern of the folded antenna with SRR on the human body is presented in Figure 8.37.

8.7 META-MATERIAL WEARABLE ANTENNAS

The proposed wearable meta-materials antennas may be placed inside a belt as shown in Figure 8.38. Three to four antennas may be placed in a belt and attached to the patient's stomach. More antennas may be attached to the patient's back to improve the level of the received signal from different locations in the human body. The cable from each antenna is connected to a recorder. The received signal is transferred via a SP8T switch to the receiver. The antennas receive a signal that is transmitted from various positions in the human body. The medical system selects the signal with the highest power.

In several systems the distance separating the transmitting and receiving antennas is in the near-field zone. In these cases the electric field intensity decays rapidly with distance. The near fields transfer energy only to close distances from the antenna and do not radiate energy to far distances. The radiated power is trapped in the region near the antenna. In the near-field zone the receiving and transmitting antennas are magnetically coupled. The inductive coupling value between two antennas is measured by their mutual inductance. In these systems we have to consider only the near-field electromagnetic coupling.

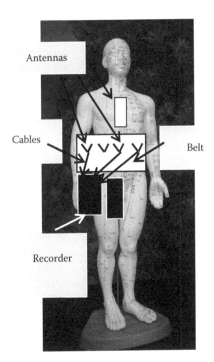

FIGURE 8.38 Medical system with printed wearable antenna.

In Figures 8.39 to 8.42 several photos of printed antennas for medical applications are shown. The dimensions of the folded dipole antenna are 7 × 6 × 0.16 cm. The dimensions of the compact folded dipole presented in Ref. [5] and shown in Figure 8.39 are 5 × 5 × 0.5 cm. The antenna's electrical characteristics on the human body have been measured by using a phantom. The phantom was designed to represent the human body electrical properties as presented in Ref. [5]. The tested antenna was attached to the phantom during the measurements of the antenna's electrical parameters.

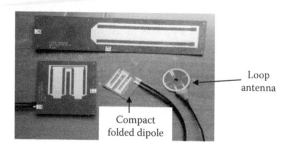

FIGURE 8.39 Microstrip antennas for medical applications.

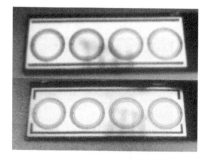

FIGURE 8.40 Meta-material antennas for medical applications.

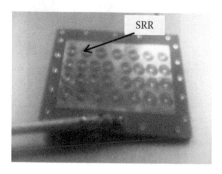

FIGURE 8.41 Meta-material patch antenna with SRR.

FIGURE 8.42   Meta-material stacked patch antenna with SRR.

8.8   WIDEBAND STACKED PATCH WITH SRR

A wideband microstrip stacked patch antenna with air spacing [1–3] has been designed. The antenna has been designed with SRR. The antenna consists of two layers. The first layer consists of a 1.6 mm thick FR4 dielectric substrate with dielectric constant of 4. The second layer consists of a 1.6 mm thick RT/Duroid 5880 dielectric substrate with dielectric constant of 2.2. The layers are separated by air spacing. The dimensions of the microstrip stacked patch antenna shown in Figure 8.43 are 33 × 20 × 3.2 mm. The antenna has been analyzed by using Agilent ADS software. The antenna bandwidth is around 10% for VSWR better than 2.0:1. The antenna beam width is around 72°. The antenna gain is around 90 dBi to 10 dBi. The antenna efficiency is around 95%. The computed S_{11} parameters are presented in Figure 8.44. The radiation pattern of the stacked patch is shown in Figure 8.45. There is a good agreement between measured and computed results.

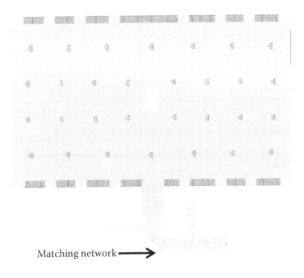

Matching network $\longrightarrow$

FIGURE 8.43   Wideband stacked patch antenna with SRR.

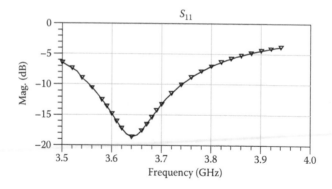

FIGURE 8.44 Wideband stacked antenna with SRR, S_{11} results.

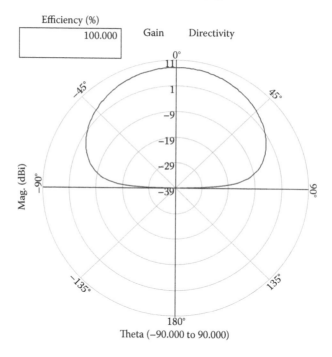

FIGURE 8.45 Radiation pattern of the stacked antenna with SRR.

Figure 8.32 presents S_{11} results of the antenna with SRR shown in Figure 8.13 on the human body. The antenna resonant frequency is shifted by 3%. Figure 8.33 presents S_{11} results of the antenna with SRR and metallic strips, shown in Figure 8.16, on the human body. The antenna resonant frequency is shifted by 1%.

8.9 SMALL META-MATERIAL ANTENNA ANALYSIS

By employing the analysis of small antennas given in Ref. [26] we can calculate the bandwidth of small antennas loaded with SRR. A small antenna is defined as an antenna whose maximum dimension is less than $\dfrac{\lambda}{2\pi}$ or $\dfrac{\lambda}{2\pi} a < 1$. The free space wavelength is λ. The radius of sphere enclosing the maximum dimension of the antenna is a. For an electrically small antenna, contained within a given volume, the antenna has a minimum value of Q [26]. This places a limit on the attainable impedance bandwidth of an electrically small antenna (ESA). The minimum Q for an electrically small linear antenna in free space is expressed as

$$Q = \frac{1}{k^3 a^3} + \frac{1}{ka} \tag{8.1}$$

The gain of a small antenna is bounded and is given as

$$G = k^2 a^2 + 2ka \tag{8.2}$$

Equation 8.2 may be solved as a quadratic equation where a represents an unknown for a given gain G.

$$0 = k^2 a^2 + 2ka - G \tag{8.3}$$

The solution of Equation 8.3 is given by Equation 8.4:

$$a = \frac{-1 + \sqrt{1 + G}}{2\pi} \lambda \tag{8.4}$$

For an efficient small meta-material antenna the effective antenna area is greater than the antenna physical area and may be written as a_M. The gain of the efficient small antenna may be written as $G_M = G\alpha_M > G$.

$$a_M = \frac{-1 + \sqrt{1 + G_M}}{2\pi} \lambda \tag{8.5}$$

Q_{LM} for a small efficient antenna may be expressed as

$$Q_{LM} = \frac{1}{k^3 a_M^3} + \frac{1}{ka_M} \tag{8.6}$$

For a given VSWR S the antenna bandwidth BW may be written as

$$BW = \frac{S-1}{Q_{LM}\sqrt{S}} \tag{8.7}$$

Q_{LM} can be expressed as

$$Q_{LM} = \frac{S-1}{BW\sqrt{S}} \tag{8.8}$$

TABLE 8.3
Computed Bandwidth of Small Antennas

Frequency (MHz)	K (rad/m)	a (cm)	ka	G (dB)	Q_{LM}	BW (%)
400	8.37	10	0.837	3.75	2.90	24.37
450	9.42	8	0.754	3.2	3.66	19.31

TABLE 8.4
Computed Bandwidth of Small Antennas with SRR

Frequency (MHz)	ka (rad/m)	a_M (cm)	ka_M	G (dB)	Q_{LM}	BW (%)
400	8.37	12.40	1.04	5	1.85	38.21
300	6.28	16.55	1.04	5	1.85	38.21

Computed bandwidth values of small antennas are given in Table 8.3. Computed values of small antennas with SRRs are given in Table 8.4.

8.10 CONCLUSION

Meta-material technology is used to develop small antennas with high efficiency. A new class of printed meta-material antennas with high efficiency is presented in this chapter. The bandwidth of the antenna with SRR and metallic strips is around 50% for VSWR better than 2.3:1. Optimization of the number of coupling stubs and the distance between the coupling stubs may be used to tune the antenna resonant frequency and number of resonant frequencies. The length of the antennas with SRR is smaller by 5% than that of the antennas without SRR. Moreover, the resonant frequency of the antennas with SRR is lower by 5%–10% than that of the antennas without SRR. The gain and directivity of the patch antenna with SRR is higher by 2–3 dB than the patch antenna without SRR. The resonant frequency of the antenna with SRR on human body is shifted by 3%.

REFERENCES

1. James, J. R., Hall, P. S., and Wood, C. *Microstrip Antenna Theory and Design*. London: The Institution of Engineering and Technology, 1981.
2. Sabban, J. A., and Gupta, K. C. Characterization of Radiation Loss from Microstrip Discontinuities Using a Multiport Network Modeling Approach. *IEEE Transactions on Microwave Theory and Techniques*, 39(4):705–712, 1991.
3. Sabban, A. A New Wideband Stacked Microstrip Antenna. In *IEEE Antenna and Propagation Symposium*, Houston, TX, June 1983.
4. Sabban, A. Microstrip Antenna Arrays. In N. Nasimuddin (Ed.), *Microstrip Antennas*, pp. 361–384, 2011. InTech, http://www.intechopen.com/articles/show/title/microstrip-antenna-arrays.
5. Sabban, A. New Wideband Printed Antennas for Medical Applications. *IEEE Transactions on Antennas and Propagation*, 61(1):84–91, 2013.

6. Pendry, J. B., Holden, A. J., Stewart, W. J., and Youngs, I. Extremely Low Frequency Plasmons in Metallic Mesostructures. *Physics Review Letters*, 76:4773–4776, 1996.

7. Pendry, J. B., Holden, A. J., Robbins, D. J., and Stewart, W. J. Magnetism from Conductors and Enhanced Nonlinear Phenomena. *IEEE Transactions on Microwave Theory and Techniques*, 47:2075–2084, 1999.

8. Marqués, R., Mesa, F., Martel, J., and Medina, F. Comparative Analysis of Edge and Broadside Coupled Split Ring Resonators for Metamaterial Design: Theory and Experiment. *IEEE Transactions on Antennas and Propagation*, 51:2572–2581, 2003.

9. Marqués, R., Baena, J. D., Martel, J., Medina, F., Falcone, F., Sorolla, M., and Martin, F. Novel Small Resonant Electromagnetic Particles for Metamaterial and Filter Design. In *Proceedings of ICEAA'03*, Torino, Italy, pp. 439–442, 2003.

10. Marqués, R., Martel, J., Mesa, F., and Medina, F. Left-Handed-Media Simulation and Transmission of EM Waves in Subwavelength Split-Ring-Resonator-Loaded Metallic Waveguides. *Physics Review Letters*, 89, paper 183901, 2002.

11. Baena, J. D., Marqués, R., Martel, J., and Medina, F. Experimental Results on Metamaterial Simulation Using SRR-Loaded Waveguides. In *Proceedings of IEEE-AP/S International Symposium on Antennas and Propagation*, Ohio, pp. 106–109, June 2003.

12. Marqués, R., Martel, J., Mesa, F., and Medina, F. A New 2-D Isotropic Left-Handed Metamaterial Design: Theory and Experiment. *Microwave and Optical Technology Letters*, 35:405–408, 2002.

13. Shelby, R. A., Smith, D. R., Nemat-Nasser, S. C., and Schultz, S. Microwave Transmission Through a Two-Dimensional, Isotropic, Left-Handed Metamaterial. *Applied Physics Letters*, 78:489–491, 2001.

14. Zhu, J., and Eleftheriades, G. V. A Compact Transmission-Line Metamaterial Antenna with Extended Bandwidth. *IEEE Antennas and Wireless Propagation Letters*, 8:295–298, 2009.

15. Chirwa, L. C., Hammond, P. A., Roy, S., and Cumming, D. R. S. Electromagnetic Radiation from Ingested Sources in the Human Intestine Between 150 MHz and 1.2 GHz. *IEEE Transactions on Biomedical Engineering*, 50(4):484–492, 2003.

16. Werber, D., Schwentner, A., and Biebl, E. M. Investigation of RF Transmission Properties of Human Tissues. *Advances in Radio Science*, 4:357–360, 2006.

17. Gupta, B., Sankaralingam S., and Dhar, S. Development of Wearable and Implantable Antennas in the Last Decade. In *Microwave Mediterranean Symposium (MMS)*, Cyprus, pp. 251–267, August 2010.

18. Thalmann, T., Popovic, Z., Notaros, B.M., and Mosig, J.R. Investigation and Design of a Multi-band Wearable Antenna. In *3rd European Conference on Antennas and Propagation, EuCAP*, Berlin, Germany, pp. 462–465, March 2009.

19. Salonen, P., Rahmat-Samii, Y., and Kivikoski, M. Wearable Antennas in the Vicinity of Human Body. In *IEEE Antennas and Propagation Society International Symposium*, Monterey, California, vol. 1, pp. 467–470, 2004.

20. Kellomaki, T., Heikkinen, J., and Kivikoski, M. Wearable Antennas for FM Reception. In *First European Conference on Antennas and Propagation, EuCAP2006*, Nice, France, pp. 1–6, November 2006.

21. Sabban, A. Wideband Printed Antennas for Medical Applications. In *APMC 2009 Conference*, Singapore, December 2009.

22. Alomainy, A., Sani, A. et al. Transient Characteristics of Wearable Antennas and Radio Propagation Channels for Ultrawideband Body-Centric Wireless Communication. *IEEE Transactions on Antennas and Propagation*, 57(4):875–884, 2009.

23. Klemm, M., and Troester, G. Textile UWB Antenna for Wireless Body Area Networks. *IEEE Transactions on Antennas and Propagation*, 54(11):3192–3197, 2006.

24. Izdebski, P. M., Rajagoplan, H., and Rahmat-Sami, Y. Conformal Ingestible Capsule Antenna: A Novel Chandelier Meandered Design. *IEEE Transactions on Antennas and Propagation*, 57(4):900–909, 2009.
25. Sabban, A. Wideband Tunable Printed Antennas for Medical Applications. *IEEE Antenna and Propagation Symposium*, Chicago, July 2012.
26. Mclean, J. S. A Reexamination of the Fundamental Limits of the Radiation Q of the Electrically Small Antennas. *IEEE Transactions on Antennas and Propagation*, 44(5): 672–675, 1996.

9 Fractal Printed Antennas

9.1 INTRODUCTION

A fractal antenna is an antenna that uses antenna design with similar fractal segments to maximize the antenna effective area. Fractal antennas are also referred to as multilevel structures with space filling curves. The key aspect lies in a repetition of a motif over two or more scale sizes or iterations. Fractal antennas are very compact, multiband, or wideband, and have useful applications in cellular telephone and microwave communications. Several fractal antennas were described in books, papers, and patents [1–15].

9.2 FRACTAL STRUCTURES

A curve, with endpoints, is represented by a continuous function whose domain is the unit interval [0, 1]. The curve may line in a plane or in a 3D space. A fractal curve is a densely self-intersecting curve that passes through every point of the unit square. A fractal curve is a continuous mapping from the unit interval to the unit square.

In mathematics, a space-filling curve (SFC) is a curve whose range contains the entire two-dimensional unit square. Most SFCs are constructed iteratively as a limit of a sequence of piecewise linear continuous curves, each one closely approximating the space-filling limit. In SFCs, where two subcurves intersect (in the technical sense), there is self-contact without self-crossing. A SFC can be (everywhere) self-crossing if its approximation curves are self-crossing. A SFC's approximations can be self-avoiding, as presented in Figure 9.1. In three dimensions, self-avoiding approximation curves can even contain joined ends. SFCs are special cases of fractal constructions. No differentiable SFC can exist.

The term "fractal curve" was introduced by B. Mandelbrot [1,2] to describe a family of geometrical objects that are not defined in standard Euclidean geometry. Fractals are geometric shapes that repeat themselves over a variety of scale sizes. One of the key properties of a fractal curve is the self-similarity. A self-similar object is unchanged after increasing or shrinking its size. An example of a repetitive geometry is the Koch curve, presented in 1904 by Helge von Koch, and is shown in Figure 9.1b. Koch generated the geometry by using a segment of straight line and raises an equilateral triangle over its middle third. The result of repeating once more the process of erecting equilateral triangles over the middle thirds of straight lines is shown in Figure 9.2a. Iterating the process infinitely many times results in a "curve" of infinite length. This geometry is continuous everywhere but is nowhere differentiable. An equilateral triangle to which the Koch process is applied will, after many iterations, converge to the Koch snowflake shown in Figure 9.2. This process can be

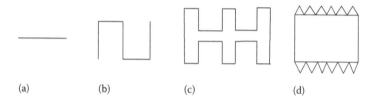

FIGURE 9.1   (a) Line. (b) Motif of bended line. (c) Bended line fractal structures. (d) Fractal structure.

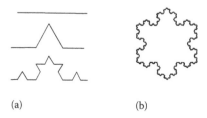

FIGURE 9.2   (a) Koch fractal structures. (b) Koch snowflakes.

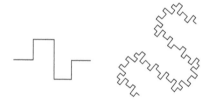

FIGURE 9.3   Folded fractal structures.

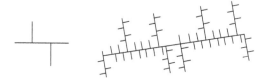

FIGURE 9.4   Variations of Koch fractal structures.

applied to several geometries as shown in Figures 9.3 and 9.4. Many variations of these geometries are presented in several articles.

9.3   FRACTAL ANTENNAS

Fractal geometries may be applied to design antennas and antenna arrays. The advantages of printed circuit technology and printed antennas enhance the design of fractal printed antennas and microwave components. The effective area of a fractal antenna is significantly higher than the effective area of a regular printed antenna.

Fractal antennas may operate with good performance at several different frequencies simultaneously. Fractal antennas are compact multiband antennas. The directivity of fractal antennas is usually higher than the directivity of a regular printed antenna. The number of elements in a fractal antenna array may be reduced by around a quarter of the number of elements in a regular array. A fractal antenna could be considered as a nonuniform distribution of radiating elements. Each of the elements contributes to the total radiated power density at a given point with a given amplitude and phase. By spatially superposing these line radiators we can study the properties of a fractal antenna array.

Small antenna main features are listed in the next paragraph:

- A large input reactance (either capacitive or inductive) that usually has to be compensated with an external matching network
- A small radiating resistance
- Small bandwidth and low efficiency
- This means that it is highly challenging to design a resonant antenna in a small space in terms of the wavelength at resonance

The use of microstrip antennas is well known in mobile telephony handsets [5]. Planar inverted-F antennas (PIFAs) configuration is popular in mobile communication systems. The advantages of PIFA antennas are their low profile, their low fabrication costs, and an easy integration within the system structure. One of the miniaturization techniques used in this antenna system is based on space-filling curves. In some particular cases of antenna configuration system, the antenna shape may be described as a multilevel structure. The multilevel technique already has been proposed to reduce the physical dimensions of microstrip antennas. The present integrated multiservice antenna system for communication system comprises the following parts and features.

The antenna includes a conducting strip or wire shaped by a SFC, composed by at least 200 connected segments forming a substantial right angle with each adjacent segment smaller than a hundredth of the free-space operating wavelength. The important reduction size of such an antenna system is obtained by using space-filling geometries. An SFC can be described as a curve that is large in terms of physical length but small in terms of the area in which the curve can be included. An SFC can be fitted over a flat or curved surface, and due to the angles between segments, the physical length of the curve is always larger than that of any straight line that can be fitted in the same area (surface).

In addition, to properly shape the structure of a miniature antenna, the segments of the SFCs must be shorter than a tenth of the free-space operating wavelength. The antenna is fed with a two-conductor structure such as a coaxial cable, with one of the conductors connected to the lower tip of the multilevel structure and the other conductor connected to the metallic structure of the system which acts as a ground plane. This antenna type features a size reduction of 20% from the typical size of a conventional external quarter-wave whip antenna. This feature, together with the small profile of the antenna which can be printed in a low-cost dielectric substrate, allows a simple and compact integration of the antenna structure.

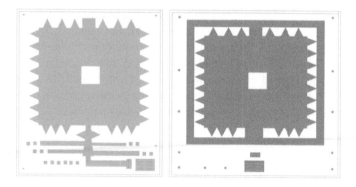

FIGURE 9.5   (a) Patch with space-filling perimeter of the conducting sheet. (b) Microstrip patch with space-filling perimeter of the conducting sheet.

Reducing the size of the radiating elements can be achieved by using a PIFA configuration, consisting of connecting two parallel conducting sheets, separated either by air or a dielectric, magnetic, or magneto dielectric material. The sheets are connected through a conducting strip near one of the sheet corners and orthogonally mounted to both sheets.

The antenna is fed through a coaxial cable, having its outer conductor connected to first sheet, being the second sheet coupled either by direct contact or capacitive to inner conductor of the coaxial cable.

In Figure 9.5a and b are presented two examples of a space-filling perimeter of the conducting sheet to achieve an optimized miniaturization of the antenna.

9.4   ANTIRADAR FRACTALS AND/OR MULTILEVEL CHAFF DISPERSERS

9.4.1   DEFINITION OF CHAFF

Chaff was one of the forms of countermeasure employed against radar. It usually consists of a large number of electromagnetic dispersers and reflectors, normally arranged in the form of strips of metal foil packed in a bundle. Chaff is usually employed to foil or to confuse surveillance and tracking radar.

9.4.2   GEOMETRY OF DISPERSERS

New geometries of the dispersers or reflectors that improve the properties of radar chaff are described in this section [6]. Some of the geometries presented here of the dispersers or reflectors are related with some forms for antennas. Multilevel and fractal structures antennas are distinguished in being of reduced size and having a multiband behavior, as has been expounded already in patent publications [7].

The main electrical characteristic of a radar chaff disperser is its radar cross section (RCS), which is related with the reflective capability of the disperser.

A fractal curve for a chaff disperser is defined as a curve comprising at least 10 segments that are connected so that each element forms an angle with its neighbors, and no pair of these segments defines a longer straight segment, as these segments are smaller than a tenth part of the resonant wavelength in free space of the entire structure of the dispenser.

In many of the configurations presented, the size of the entire disperser is smaller than a quarter of the lowest operating wavelength.

The SFCs (or fractal curves) can be characterized as follows:

1. They are long in terms of physical length but small in terms of area in which the curve can be included. The dispersers with a fractal form are long electrically but can be included in a very small surface area. This means it is possible to obtain a smaller packaging and a denser chaff cloud using this technique.
2. Frequency response: Their complex geometry provides a spectrally richer signature when compared with rectilinear dispersers known in the state of the art.

The fractal structure properties of disperser introduce an advantage not only in terms of reflected radar signal response, but also in terms of aerodynamic profile of dispersers. It is known that a surface offers greater resistance to air than a line or a one-dimensional form.

Therefore, giving a fractal form to the dispersers with a dimension greater than unity ($D > 1$) increases resistance to the air and improves the time of suspension.

9.5 DEFINITION OF MULTILEVEL STRUCTURE

Multilevel structures are a geometry related with fractal structures. In that case of radar chaff a multilevel structure is defined as structure that includes a set of polygons, which are characterized in having the same number of sides, wherein these polygons are electromagnetically coupled either by means of capacitive coupling, or by means of an ohmic contact. The region of contact between the directly connected polygons is smaller than 50% of the perimeter of the polygons mentioned in at least 75% of the polygons that constitute the defined multilevel structure.

A multilevel structure provides both

- A reduction in the size of dispersers and an enhancement of their frequency response
- Can resonate in a nonharmonic way, and can even cover simultaneously and with the same relative bandwidth at least a portion of numerous bands

The fractal structures (SFC) are preferred when a reduction in size is required, whereas multilevel structures are preferred when it is required that the most important considerations be given to the spectral response of radar chaff.

The main advantages for configuring the form of the chaff dispersers are

1. The dispersers are small; consequently more disperser can be encapsulated in a same cartridge, rocket, or launch vehicle.

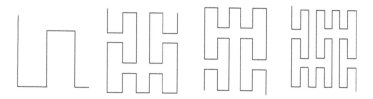

FIGURE 9.6   Fractal curves that can be used to configure a chaff disperser.

FIGURE 9.7   Hilbert fractal curves.

2. The dispersers are also lighter; therefore they can remain more time floating in the air than the conventional chaff.
3. Because of the smaller size of the chaff dispersers, the launching devices (cartridges, rockets, etc.) can be smaller with regard to chaff systems in the state of the art providing the same RCS.
4. Because of the lighter weight of the chaff dispersers, the launching devices can shoot the packages of chaff farther from the launching devices and locations.
5. Chaff constituted by multilevel and fractal structures provide larger RCS at longer wavelengths than conventional chaff dispersers of the same size.
6. The dispersers with long wavelengths can be configured and printed on light dielectric supports having a nonaerodynamic form and opposing a greater resistance to the air and thereby having a longer time of suspension.
7. The dispersers provide a better frequency response with regard to dispersers of the state of the art.

In Figure 9.6 such size compression structures based on fractal curves are presented. Figure 9.7 shows several examples of Hilbert fractal curves (with increasing iteration order) which can be used to configure the chaff disperser.

9.6   ADVANCED ANTENNA SYSTEM

The main advantage of advanced antenna system lies in the multiband and multiservice performance of the antenna. This enables convenient and easy connection of a simple antenna for most communication systems and applications. The main advantages addressed by advanced antennas featured similar parameters (input impedance, radiation pattern) at several bands maintaining their performance, compared with conventional antennas. Fractal shapes make it possible to obtain a compact antenna of reduced dimensions compared to other conventional antenna. Multilevel antennas introduced a higher flexibility to design multiservice antennas for real applications, extending the theoretical capabilities of ideal fractal antennas to practical, commercial antennas.

9.7 COMPARISON BETWEEN EUCLIDEAN AND FRACTAL ANTENNAS

Most conventional antennas are Euclidean design/geometry, where the closed antenna area is directly proportional to the antenna perimeter. Thus, for example, when the length of a Euclidean square is increased by a factor of three, the enclosed area of the antenna is increased by a factor of nine. Gain, directivity, impedance, and efficiency of Euclidean antennas are a function of the antenna's size-to-wavelength ratio.

It is typically desired that Euclidean antennas operate within a narrow range (e.g., 10%–40%) around a central frequency f_c which in turn dictates the size of the antenna (e.g., half or quarter wavelength). When the size of a Euclidean antenna is made much smaller than the operating wavelength (λ), it becomes very inefficient because the antenna's radiation resistance decreases and becomes less than its ohmic resistance (i.e., it does not couple electromagnetic excitations efficiently to free space). Instead, it stores energy reactively within its vicinity (reactive impedance X_c). These aspects of Euclidean antennas work together to make it difficult for small Euclidean antennas to couple or match to feeding or excitation circuitry and cause them to have a high Q factor (lower bandwidth). Q (Quality) factor may be defined as approximately the ratio of input reactance X_{in} to radiation resistance R_r, $Q = X_{in}/R_r$.

The Q factor may also be defined as the ratio of average stored electric energies (or magnetic energies stored) to the average radiated power. Q can be shown to be inversely proportional to bandwidth. Thus, small Euclidean antennas have very small bandwidth, which is of course undesirable (matching network may be needed). Many known Euclidean antennas are based on closed-loop shapes.

Unfortunately, when small in size, such loop-shaped antennas are undesirable because, as discussed previously, the radiation resistance decreases significantly when the antenna size is decreased. This is because the physical area (A) contained within the loop-shaped antenna's contour is related to the loop perimeter.

Radiation resistance R_r of a circular loop-shaped Euclidean antenna is defined by R_r, as given in Equation 9.1, where k is a constant.

$$R_r = \eta\pi(2/3)(KA/\lambda)^2 = 20\pi^2(C/\lambda) \tag{9.1}$$

As the resistance R_c is only proportional to perimeter (C), then for $C < 1$, the resistance R_c is greater than the radiation resistance R_r and the antenna is highly inefficient. This is generally true for any small circular Euclidean antenna. A small-sized antenna will exhibit a relatively large ohmic resistance and a relatively small radiation resistance R_r. This low efficiency limits the use of the small antennas.

Fractal geometry is a non-Euclidean geometry that can be used to overcome the problems with small Euclidean antennas. Radiation resistance R_r of a fractal antenna decreases as a small power of the perimeter (C) compression, with a fractal loop or island always having a substantially higher radiation resistance than a small Euclidean loop antenna of equal size. Fractal geometry may be grouped into

- Random fractals, which may be called chaotic or Brownian fractals.
- Deterministic or exact fractals. In deterministic fractal geometry, a self-similar structure results from the repetition of a design or motif (generator) with self-similarity and structure at all scales. In deterministic or exact self-similarity, fractal antennas may be constructed through recursive or iterative means. In other words, fractals are often composed of many copies of themselves at different scales, thereby allowing them to defy the classical antenna performance constraint, which is size-to-wavelength ratio.

9.8   MULTILEVEL AND SPACE-FILLING GROUND PLANES FOR MINIATURE AND MULTIBAND ANTENNAS

A new family of antenna ground planes of reduced size and enhanced performance is based on an innovative set of geometries. These new geometries are known as multilevel and space-filling structures, which had been previously used in the design of multiband and miniature antennas.

One of the key issues of the present antenna system is considering the ground plane of an antenna as an integral part of the antenna that mainly contributes to its radiation and impedance performance (impedance level, resonant frequency, and bandwidth).

The multilevel and space-filling structures are used in the ground plane of the antenna, in this way allowing a better return loss or voltage standing wave ratio (VSWR), a better bandwidth, a multiband behavior, or a combination of all these effects. The technique can be seen as well as a means of reducing the size of the ground plane and therefore the size of the overall antenna. The key point of the present antenna system is shaping the ground plane of an antenna in such a way that the combined effect of the ground plane and the radiating element enhances the performance and characteristics of the whole antenna device, in terms of bandwidth, VSWR, multiband, efficiency, size, or gain.

9.8.1   MULTILEVEL GEOMETRY

The resulting geometry is no longer a solid, conventional ground plane, but a ground plane with a multilevel or space-filling geometry, at least in a portion of it.

A multilevel geometry for a ground plane consists of a conducting structure including a set of polygons, featuring the same number of sides, electromagnetically coupled by means of either a capacitive coupling or ohmic contact. The contact region between directly connected polygons is narrower than 50% of the perimeter of polygons in at least 75% of polygons defining the conducting ground plane. In this definition of multilevel geometry, circles and ellipses are included as well, as they can be understood as polygons with an infinite number of sides.

9.8.2   SPACE-FILLING CURVE

An SFC is a curve that is large in terms of physical length but small in terms of the area in which the curve can be included.

A straight longer segment is defined as follows. Here a curve is composed by at least 10 segments that are connected in such a way that each segment forms an angle with its neighbors, that is, no pair of adjacent segments define a larger straight segment. The curve can be optionally periodic along a fixed straight direction of space if, and only if, the period is defined by a nonperiodic curve composed by at least 10 connected segments and no pair of adjacent and connected segments.

An SFC can be fitted over a flat or curved surface, and because of the angles between segments, the physical length of the curve is always larger than that of any straight line that can be fitted in the same area (surface). In addition, to properly shape the ground plane, the segments of the SFC curves included in the ground plane must be shorter than a tenth of the free-space operating wavelength.

Figure 9.8 shows several examples of fractal geometries that can be used as SFCs. Figure 9.9 shows several examples of Hilbert fractal curves that can be used as SFCs.

The curves shown in Figure 9.8 are some examples of such SFCs. Owing to the special geometry of multilevel and space-filling structure, the current distributes over the ground plane in such a way that it enhances the antenna performance and features in terms of

- Reduced size compared to antennas with a solid ground plane
- Enhanced bandwidth compared to antennas with a solid ground plane
- Multifrequency performance
- Better VSWR feature at the operating band or bands
- Better radiation efficiency
- Enhanced gain

Figure 9.10a shows a patch antenna above a particular example of a new ground-plane structure formed by both multilevel and space-filling geometries. Figure 9.10b shows a monopole antenna above a ground-plane structure formed by both multilevel and space-filling geometries. Figure 9.11 shows several examples of different contour shaped for multilevel ground planes, such as rectangular (Figure 9.11a and b) and circular ground plane (Figure 9.11c).

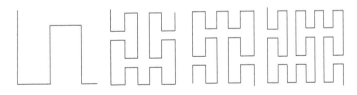

FIGURE 9.8   Fractal curves that can be used as space-filling curves.

FIGURE 9.9   Hilbert fractal curves that can be used as space-filling curves.

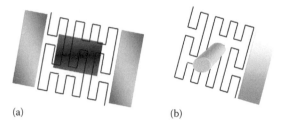

FIGURE 9.10 (a) Patch antenna above a new ground-plane structure. (b) Monopole antenna above a ground-plane structure formed by both multilevel and space-filling geometries.

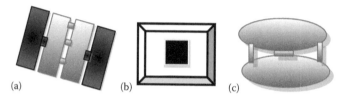

FIGURE 9.11 Examples of different contour shaped multilevel ground planes. (a) Rectangular ground planes. (b) Multilevel rectangular ground plane. (c) Circular ground planes.

9.9 APPLICATIONS OF FRACTAL PRINTED ANTENNAS

In this chapter several designs of fractal printed antennas are presented for communication applications. These fractal antennas are compact and efficient. The antenna gain is around 8 dBi with 90% efficiency.

9.9.1 NEW 2.5-GHZ FRACTAL PRINTED ANTENNAS WITH SPACE-FILLING PERIMETER ON THE RADIATOR

A new fractal microstrip antenna was designed as presented in Figure 9.12. The antenna was printed on a 0.8 mm thick Duroid® substrate with 2.2 dielectric constant. The antenna dimensions are 5.2 × 48.8 × 0.08 cm. The antenna was designed by using ADS software. The antenna bandwidth is around 2% around 2.5 GHz for VSWR better than 3:1. The antenna bandwidth may be improved to 5%, for VSWR

FIGURE 9.12 Fractal antenna resonators.

better than 2:1, by adding a second layer above the resonator. A patch radiator is printed on the second layer as presented in Figure 9.13. The radiator was printed on a 0.8 mm thick FR4 substrate with 4.5 dielectric constant. The electromagnetic fields radiated by the resonator are electromagnetic coupled to the patch radiator. The patch radiator dimensions are $45.2 \times 48.8 \times 0.08$ cm. The stacked fractal antenna structure is shown in Figure 9.14. The spacing between the two layers may be varied to get a wider bandwidth. The stacked fractal antenna S_{11} parameters with 8 mm air spacing between the layers is presented in Figure 9.15. The S_{11} parameter of the

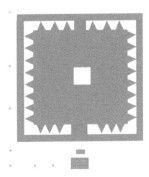

FIGURE 9.13   Fractal antenna patch radiator.

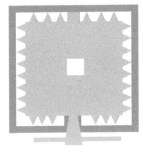

FIGURE 9.14   Fractal stacked patch antenna structure.

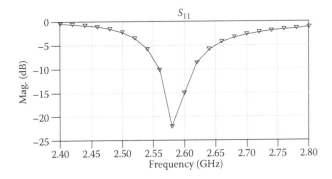

FIGURE 9.15 S_{11} parameter of the fractal stacked patch antenna with 8 mm air spacing.

fractal stacked patch antenna with 10 mm air spacing is given in Figure 9.16. The antenna bandwidth is improved to 5% for VSWR better than 2:1. The fractal stacked patch antenna radiation pattern is shown in Figure 9.17. The antenna beam width is around 76°, with 8 dBi gain and 91% efficiency.

A photo of the stacked fractal patch antenna is shown in Figure 9.18. The antenna resonator is shown in Figure 9.18a. The antenna radiator is shown in Figure 9.18b.

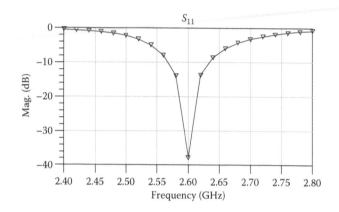

FIGURE 9.16 S_{11} parameter of the fractal stacked patch antenna with 10 mm air spacing.

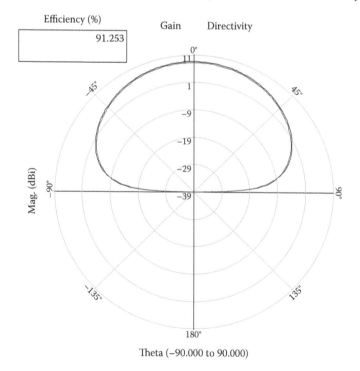

FIGURE 9.17   Fractal stacked patch antenna radiation pattern with 10 mm air spacing.

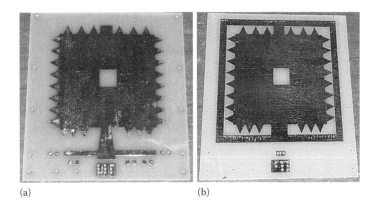

FIGURE 9.18 Fractal stacked patch antenna. (a) Resonator. (b) Radiator.

A modified version of the antenna is shown in Figure 9.19. The S_{11} parameter of the modified fractal stacked patch antenna with 8 mm air spacing is given in Figure 9.20. The antenna bandwidth is around 10% for VSWR better than 3:1. The fractal stacked patch antenna radiation pattern is shown in Figure 9.21. The antenna beam width is around 76°, with 8 dBi gain and 91.82% efficiency.

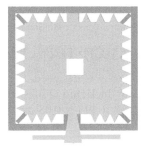

FIGURE 9.19 A modified fractal stacked patch antenna structure.

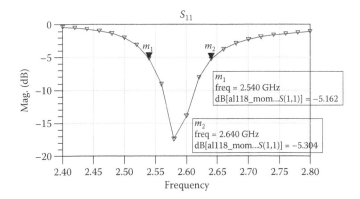

FIGURE 9.20 S_{11} parameter of the modified fractal stacked patch antenna with 8 mm air spacing.

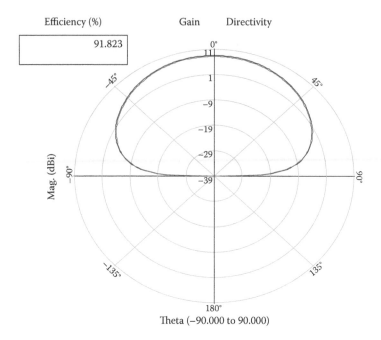

FIGURE 9.21   Fractal stacked patch antenna radiation pattern with 8 mm air spacing.

9.9.2 NEW STACKED PATCH 2.5-GHz FRACTAL PRINTED ANTENNAS

A new fractal microstrip antenna was designed as presented in Figure 9.22. The antenna was printed on a 0.8 mm thick Duroid substrate with 2.2 dielectric constant. The antenna dimensions are $45.8 \times 39.1 \times 0.08$ cm. The antenna was designed by using ADS software.

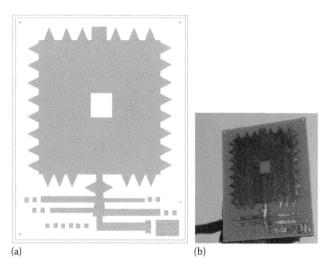

(a)   (b)

FIGURE 9.22   Resonator of a fractal stacked patch antenna. (a) Layout. (b) Resonator.

The antenna resonator bandwidth is around 2% around 2.52 GHz for VSWR better than 3:1. The antenna bandwidth may be improved to 6%, for VSWR better than 3:1, by adding a second layer above the resonator. A patch radiator is printed on the second layer as presented in Figure 9.23. The radiator was printed on a 0.8 mm thick FR4 substrate with 4.5 dielectric constant. The electromagnetic fields radiated by the resonator are electromagnetic coupled to the patch radiator. The patch radiator dimensions are 45.8 × 39.1 × 0.08 cm. The stacked fractal antenna structure is shown in Figure 9.24. The spacing between the two layers may be varied to get wider bandwidth. The single-layer fractal antenna S_{11} parameters are presented in Figure 9.25.

A comparison of computed and measured S_{11} parameters of the fractal stacked patch antenna with no air spacing is given in Figure 9.25. There is a good agreement of measured and computed results. The fractal stacked patch antenna radiation pattern is shown in Figure 9.26. The antenna beam width is around 82°, with 7.5 dBi gain and 97.2% efficiency. A photo of the fractal stacked patch antenna is shown in Figure 9.27. The antenna resonator is shown in Figure 9.28a. The antenna radiator is shown in Figure 9.28b.

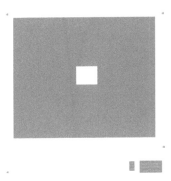

FIGURE 9.23   Radiator of a fractal stacked patch antenna.

FIGURE 9.24   Layout of the fractal stacked patch antenna.

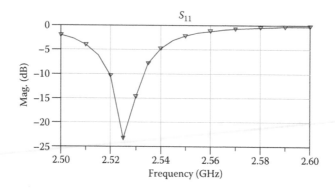

FIGURE 9.25 Computed S_{11} parameter of the single-layer fractal antenna.

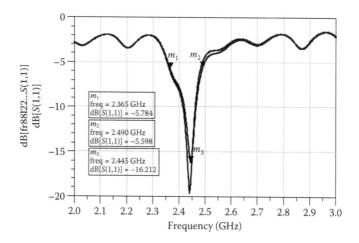

FIGURE 9.26 Measured and computed S_{11} parameters of the fractal stacked patch antenna with no air spacing between the layers.

9.9.3 NEW 8-GHZ FRACTAL PRINTED ANTENNAS WITH SPACE-FILLING PERIMETER OF THE CONDUCTING SHEET

A new fractal microstrip antenna was designed as presented in Figure 9.29. The antenna was printed on a 0.8 mm thick Duroid substrate with 2.2 dielectric constant. The antenna dimensions are 17.2 × 21.8 × 0.08 cm. The antenna was designed by using ADS software.

The antenna resonator bandwidth is around 3% around 7.2 GHz for VSWR better than 2:1. The antenna bandwidth is improved to 22%, for VSWR better than 3:1, by adding a second layer above the resonator. A patch radiator is printed on the second layer as presented in Figure 9.30. The radiator was printed on a 0.8 mm thick FR4 substrate with 4.5 dielectric constant. The electromagnetic fields radiated by the resonator are electromagnetically coupled to the patch radiator. The patch radiator

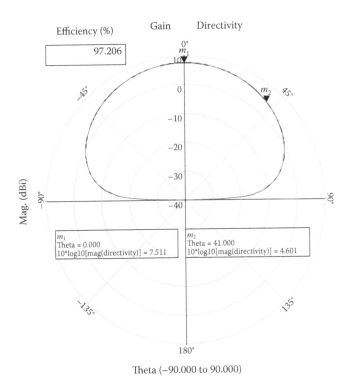

Efficiency (%)   Gain   Directivity

97.206

m_1
Theta = 0.000
10*log10[mag(directivity)] = 7.511

m_2
Theta = 41.000
10*log10[mag(directivity)] = 4.601

Theta (−90.000 to 90.000)

FIGURE 9.27   Fractal stacked patch antenna radiation pattern with 8 mm air spacing.

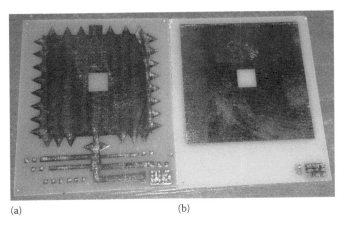

(a)   (b)

FIGURE 9.28   Fractal stacked patch antenna. (a) Resonator. (b) Radiator.

dimensions are $17.2 \times 21.8 \times 0.08$ cm. The stacked fractal antenna structure is shown in Figure 9.31. The spacing between the two layers may be varied to get wider bandwidth. The single-layer fractal antenna S_{11} parameters are presented in Figure 9.32. The computed S_{11} parameter of the fractal stacked patch antenna with 2 mm air spacing is given in Figure 9.33. The fractal stacked patch antenna radiation pattern at 7.5

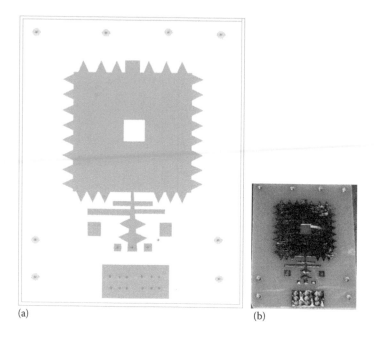

(a)   (b)

FIGURE 9.29   (a) Resonator of the 8-GHz fractal stacked patch antenna. (b) 8-GHz fractal resonator.

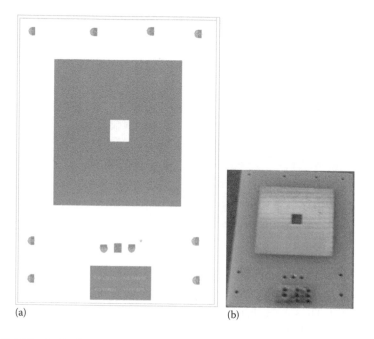

(a)   (b)

FIGURE 9.30   (a) Radiator of the 8-GHz fractal stacked patch antenna. (b) Radiator.

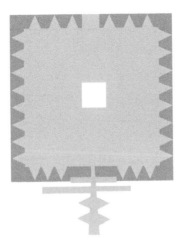

FIGURE 9.31 Layout of the 8-GHz fractal stacked patch antenna.

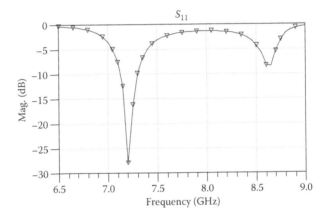

FIGURE 9.32 Computed S_{11} of the 8-GHz fractal stacked patch antenna.

GHz is shown in Figure 9.34. The antenna beam width is around 82°, with 7.8 dBi gain and 82.2% efficiency. The fractal stacked patch antenna radiation pattern at 8 GHz is shown in Figure 9.35. The antenna beam width is around 82°, with 7.5 dBi gain and 95.3% efficiency. A photo of the fractal stacked patch antenna is shown in Figure 9.36. The antenna resonator is shown in Figure 9.36a. The antenna radiator is shown in Figure 9.36b.

9.9.4 New Stacked Patch 7.4-GHz Fractal Printed Antennas

A new fractal microstrip antenna was designed as presented in Figure 9.37. The antenna was printed on a 0.8 mm thick Duroid substrate with 2.2 dielectric constant.

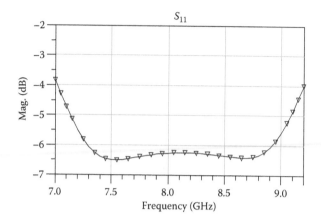

FIGURE 9.33   Computed S_{11} of the 8-GHz fractal stacked patch antenna with 2 mm air spacing.

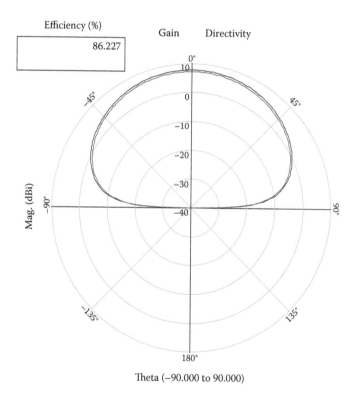

FIGURE 9.34   Fractal stacked patch antenna radiation pattern with 2 mm air spacing at 7.5 GHz.

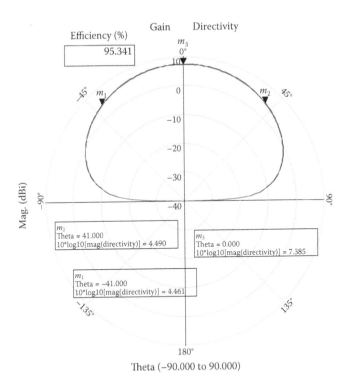

Efficiency (%)

Gain   Directivity

95.341

m_2
Theta = 41.000
10*log10[mag(directivity)] = 4.490

m_3
Theta = 0.000
10*log10[mag(directivity)] = 7.385

m_1
Theta = −41.000
10*log10[mag(directivity)] = 4.461

Mag. (dBi)

Theta (−90.000 to 90.000)

FIGURE 9.35 Fractal stacked patch antenna radiation pattern with 2 mm air spacing at 8 GHz.

(a)   (b)

FIGURE 9.36 A fractal stacked patch antenna. (a) Resonator. (b) Radiator.

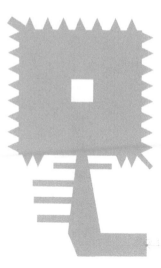

FIGURE 9.37   Layout of the 7.4-GHz fractal resonator.

The antenna dimensions are $18 \times 12 \times 0.08$ cm. The antenna was designed by using ADS software.

The antenna resonator bandwidth is around 2% around 7.4 GHz for VSWR better than 2:1. The antenna bandwidth is improved to 10%, for VSWR better than 3:1, by adding a second layer above the resonator. A patch radiator is printed on the second layer as presented in Figure 9.38. The radiator was printed on a 0.8 mm thick FR4 substrate with 4.5 dielectric constant. The electromagnetic fields radiated by the resonator are electromagnetically coupled to the patch radiator. The fractal stacked patch dimensions are $18 \times 12 \times 0.08$ cm. The stacked fractal antenna structure is shown in Figure 9.39. The spacing between the two layers may be varied to get wider bandwidth. The computed S_{11} parameter of the fractal stacked patch antenna with

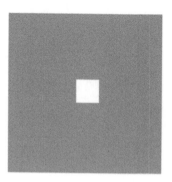

FIGURE 9.38   Radiator of a 7.4-GHz fractal stacked patch antenna.

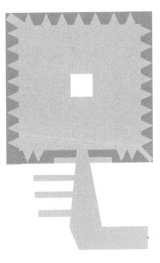

FIGURE 9.39 Layout of the 8-GHz fractal stacked patch antenna.

3 mm air spacing is given in Figure 9.40. The fractal stacked patch antenna radiation pattern at 7.5 GHz is shown in Figure 9.41. The antenna beam width is around 86°, with 7.9 dBi gain and 89.7% efficiency.

A modified antenna structure is presented in Figure 9.42. The antenna matching network has been modified and S_{11} at 7.45 GHz is –23.5 dB as shown in Figure 9.43. The computed S_{11} parameter of the fractal stacked patch antenna with 3 mm air spacing is given in Figure 9.43. The fractal stacked patch antenna bandwidth is around 9% for VSWR better than 3:1. The fractal stacked patch antenna radiation pattern at 7.5 GHz is shown in Figure 9.44. The antenna beamwidth is around 85°, with 7.8 dBi gain and 86.2% efficiency.

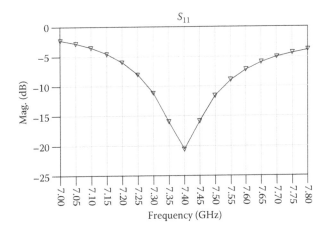

FIGURE 9.40 Computed S_{11} of the 7.4-GHz modified fractal antenna with 3 mm air spacing.

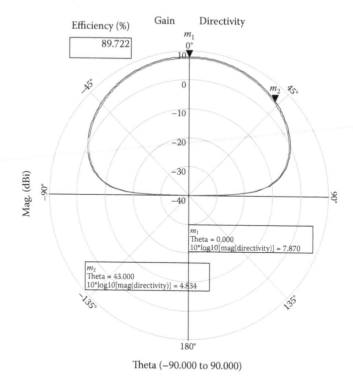

FIGURE 9.41 Fractal stacked patch antenna radiation pattern with 3 mm air spacing at 7.5 GHz.

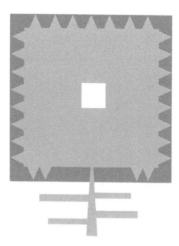

FIGURE 9.42 Layout of the modified 7.4-GHz fractal stacked patch antenna.

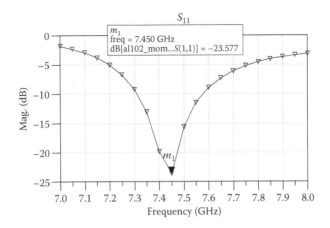

FIGURE 9.43 Computed S_{11} of the 7.4-GHz modified fractal antenna with 3 mm air spacing.

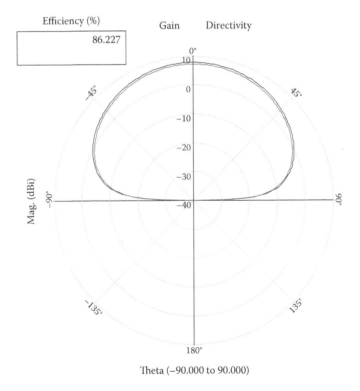

FIGURE 9.44 Modified fractal patch antenna radiation pattern with 3 mm air spacing at 7.5 GHz.

9.10   NEW FRACTAL PRINTED ANTENNAS
USING DOUBLE-LAYER HILBERT CURVES

In the next paragraphs new fractal printed antennas using double-layer Hilbert curves are presented. The antennas were designed by using ADS software.

9.10.1   NEW 3.3-GHz FRACTAL PRINTED ANTENNAS
USING DOUBLE-LAYER HILBERT CURVES

A new fractal microstrip antenna using Hilbert curves was designed as presented in Figure 9.45. The antenna was printed on a 0.8 mm thick Duroid substrate with 2.2 dielectric constant. The antenna dimensions are 55.4 × 47.3 × 0.08 cm. The antenna was designed by using ADS software.

The antenna bandwidth is around 1.5% around 3.34 GHz for VSWR better than 3:1. The antenna bandwidth may be improved to 5%, for VSWR better than 2:1, by adding a second layer above the resonator. A fractal microstrip antenna using Hilbert curves is printed on the second layer as presented in Figure 9.46. The radiator was printed on a 0.8 mm thick FR4 substrate with 4.5 dielectric constant. The electromagnetic fields radiated by the resonator are electromagnetically coupled to the patch radiator. The patch radiator dimensions are 55.4 × 47.3 × 0.08 cm. The spacing between the two layers may be varied to get wider bandwidth. The S_{11} parameters of the fractal antenna with Hilbert curves are presented in Figure 9.47. The fractal stacked patch antenna radiation pattern at 7.5 GHz is shown in Figure 7.44. The antenna behaves like a monopole antenna and has a null in the bore-sight direction. The antenna gain is 3 dBi. The antenna efficiency is 85.8%. A photo of the fractal antenna resonator with Hilbert curves is presented in Figure 9.49a. A photo of the fractal antenna resonator with Hilbert curves is presented in Figure 9.49b.

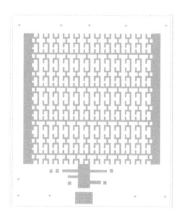

FIGURE 9.45   Fractal antenna resonator with Hilbert curves.

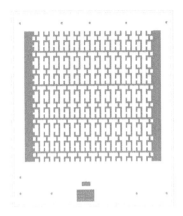

FIGURE 9.46   Radiator of a 3.3-GHz fractal stacked patch antenna with Hilbert curves.

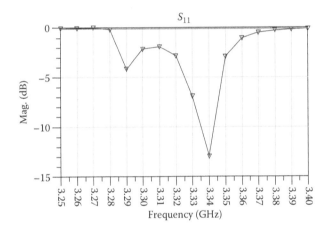

FIGURE 9.47   Computed S_{11} parameter of the single-layer fractal antenna with Hilbert curves.

9.10.2   NEW 3.3-GHZ FRACTAL PRINTED ANTENNAS USING HILBERT CURVES ON THE RESONATOR LAYER

A new fractal microstrip antenna using Hilbert curves was designed as presented in Figure 9.50. The antenna was printed on a 0.8 mm thick Duroid substrate with 2.2 dielectric constant. The antenna dimensions are $55.4 \times 47.3 \times 0.08$ cm. The antenna was designed by using ADS software.

The antenna bandwidth is around 1.5% around 3.34 GHz for VSWR better than 3:1. The antennas have two resonant frequencies on 3.35 GHz and on 3.51 GHz. The

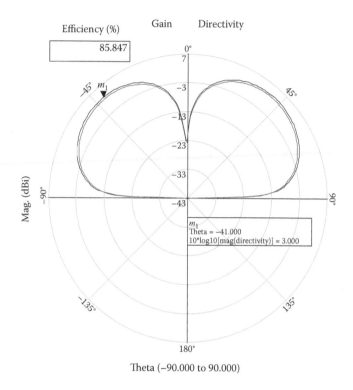

FIGURE 9.48 Computed radiation pattern of the single-layer fractal antenna with Hilbert curves.

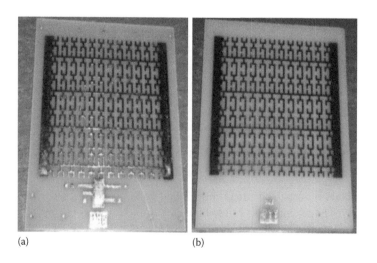

FIGURE 9.49 (a) Fractal antenna resonators. (b) Fractal antenna radiator.

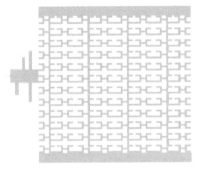

FIGURE 9.50 Fractal antenna resonator with Hilbert curves.

antenna bandwidth may be improved to 3%, for VSWR better than 3:1, by adding a second layer above the resonator. A patch microstrip antenna is printed on the second layer as shown in Figure 9.51. The radiator was printed on a 0.8 mm thick FR4 substrate with 4.5 dielectric constant. The electromagnetic fields radiated by the fractal resonator are electromagnetically coupled to the patch radiator. The patch radiator dimensions are $55.4 \times 47.3 \times 0.08$ cm. The spacing between the two layers may be varied to get a wider bandwidth. A double-layer antenna with Hilbert fractal curves on the radiator layer is shown in Figure 9.51. The S_{11} parameters of the fractal antenna with Hilbert curves on the resonator layer is presented in Figure 9.52. The fractal stacked patch antenna radiation pattern is shown in Figure 9.53. The antenna behaves like a monopole antenna and has a null in the bore-sight direction. The antenna gain is 7.8 dBi. The antenna efficiency is 71.3%. A photo of the fractal antenna resonator with Hilbert curves is presented in Figure 9.54. The gain of the antenna presented in Figure 9.51 is higher by 4.8 dB than the double-layer fractal antenna shown in Figure 9.46.

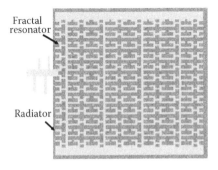

FIGURE 9.51 Double-layer fractal antenna with fractal resonator with Hilbert curves.

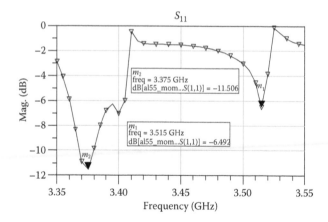

FIGURE 9.52 Computed S_{11} parameter of the single-layer fractal antenna with Hilbert curves.

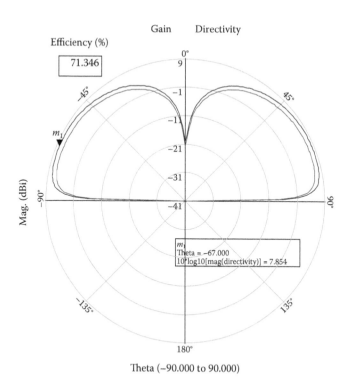

FIGURE 9.53 Computed radiation pattern of the single-layer fractal antenna with Hilbert curves.

FIGURE 9.54   Fractal antenna resonators.

9.11   CONCLUSIONS

Several designs of fractal printed antennas have been presented for communica-tion applications. These fractal antennas are compact and efficient. The space-filling technique and Hilbert curves were employed to design the fractal antennas. The antenna bandwidth is around 10% with VSWR better than 3:1. The antenna gain is around 8 dBi with 90% efficiency.

REFERENCES

1. Mandelbrot, B. B. *The Fractal Geometry of Nature.* New York: W. H. Freeman and Company, 1983.
2. Mandelbrot, B. B. How Long Is the Coast of Britain? Statistical Self-similarity and Fractional Dimension. *Science*, 156:636–638, 1967.
3. Falkoner, F. J. *The Geometry of Fractal Sets.* Cambridge, UK: Cambridge University Press, 1990.
4. Balanis, C. A. *Antenna Theory Analysis and Design*, 2nd ed. New York: John Wiley & Sons, 1997.
5. Virga, K., and Rahmat-Samii, Y. Low-Profile Enhanced-Bandwidth PIFA Antennas for Wireless Communications Packaging. *IEEE Transactions on Microwave Theory and Techniques*, 45 (10):1879–1888, October 1997.
6. Skolnik, M. I. *Introduction to Radar Systems.* London: McGraw-Hill, 1981.
7. Patent US 5087515, Patent US 4976828, Patent US 4763127, Patent US 4600642, Patent US 3952307, Patent US 3725927.
8. European Patent Application EP 1317018 A2/27.11.2002.
9. Chiou, T., and Wong, K. Design of Compact Microstrip Antennas with a Slotted Ground Plane. In *IEEE-APS Symposium*, Boston, July 8–12, 2001.
10. Hansen, R. C. Fundamental Limitations on Antennas. *Proceedings of IEEE*, 69 (2): 170–182, 1981.
11. Pozar, D. *The Analysis and Design of Microstrip Antennas and Arrays.* Piscataway, NJ: Wiley-IEEE Press, 1995.

12. Zurcher, J. F., and Gardiol, F. E. *Broadband Patch Antennas*. Norwood, MA: Artech House, 1995.

13. Minin, I. Microwave and Millimeter Wave Technologies from Photonic Bandgap Devices to Antenna and Applications, Chapter 16, *Fractal Antenna Applications*, M. V. Rusu and R. Baican (eds.), pp. 351–382, March 2010.

14. Rusu, M. V., Hirvonen, M., Rahimi, H., Enoksson, P., Rusu, C., Pesonen, N., Vermesan, O., and Rustad, H. Minkowski Fractal Microstrip Antenna for RFID Tags. In *Proceedings of the EuMW2008 Symposium*, Amsterdam, October, 2008.

15. Rahimi, H., Rusu, M., Enoksson, P., Sandström, D., and Rusu, C. Small Patch Antenna Based on Fractal Design for Wireless Sensors. MME07, In *18th Workshop on Micromachining, Micromechanics, and Microsystems*, September 16–18 2007, Portugal.

10 Microwave and MM Wave Technologies

10.1 INTRODUCTION

The communication and radar industries in microwave and mm wave frequencies are continuously growing. Radio frequency (RF) modules such as front end, filters, power amplifiers, antennas, passive components, and limiters are important modules in radar and communication links (see Refs. [1–9]). Microwave and mm wave technologies include

- Microwave integrated circuits (MICs)
- Monolithic microwave integrated circuits (MMICs)
- Micro-electro-mechanical systems (MEMSs)
- Low-temperature co-fired ceramics (LTCCs)

The electrical performance of the modules determines if the system will meet the required specifications. Moreover, in several cases the module's performance limits the system performance. Minimization of the size and weight of the RF modules is achieved by employing MMIC and MIC technology. However, integration of MIC and MMIC components and modules raises several technical challenges. Design parameters that may be neglected at low frequencies cannot be ignored in the design of wideband integrated RF modules. Powerful RF design software, such as ADS and HFSS, is required to achieve accurate design of RF modules in mm wave frequencies. Accurate design of mm wave RF modules is crucial. It is an impossible mission to tune mm wave RF modules in the fabrication process.

10.2 MICROWAVE INTEGRATED CIRCUITS

Traditional microwave systems consist of connectorized components (such as amplifiers, filters, and mixers) connected by coaxial cables. These modules have large dimensions and suffer from high losses. Dimensions and losses may be minimized by using MIC technology. There are three types of MIC circuits: hybrid microwave integrated circuit (HMIC), standard MIC, and miniature HMIC. Solid-state and passive elements are bonded to the dielectric substrate. The passive elements are fabricated by using thick or thin film technology. Standard MIC uses a single-level metallization for conductors and transmission lines. Miniature HMICs use a multilevel process in which passive elements such as capacitors and resistors are batch

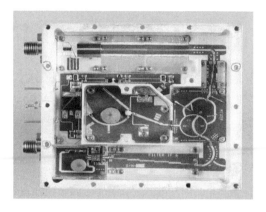

FIGURE 10.1   An MIC receiving link.

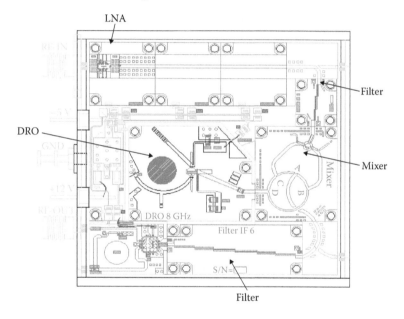

FIGURE 10.2   Layout of an MIC receiving link.

deposited on the substrate. Semiconductor devices such as amplifiers and diodes are bonded on the substrate. Figure 10.1 presents an MIC receiving link. Figure 10.2 shows the layout of the MIC receiving link. The receiving channel consists of a low-noise amplifier, filters, dielectric resonant oscillators, and a diode mixer.

10.3   MONOLITHIC MICROWAVE INTEGRATED CIRCUITS

MMICs are circuits in which active and passive elements are formed on the same dielectric substrate, as shown in Figure 10.3, by using a deposition scheme as epitaxy, ion implantation, sputtering, evaporation, and diffusion.

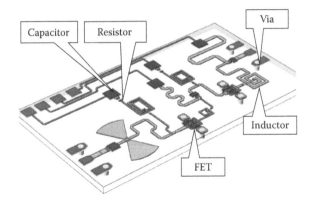

FIGURE 10.3 MMIC basic components. MMIC chip consists of passive elements such as resistors, capacitors, inductors, and a field effect transistor (FET).

10.3.1 MMIC DESIGN FACTS

MMIC components cannot be tuned. Accurate design is crucial in the design of MMIC circuits. Accurate design may be achieved by using 3D electromagnetic software.

Materials employed in the design of MMIC circuits are gallium arsenide (GaAs), indium phosphide (InP), gallium nitride (GaN), and silicon germanium (SiGe). Large statistic scattering of all electrical parameters causes sensitivity of the design.

Fabrication runs are expensive, around $200,000 per run. Miniaturization of components yields lower cost of the MMIC circuits. Figure 10.4 presents an MMIC design flow.

The designer's goal is to comply with customer specifications in one design iteration.

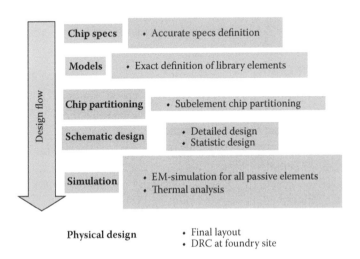

FIGURE 10.4 MMIC design flow.

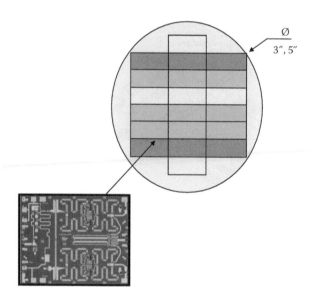

FIGURE 10.5   GaAs wafer layout.

10.3.2   MMIC TECHNOLOGY FEATURES

- 0.25 micron GaAs pseudomorphic high electron mobility transistor (PHEMT) for power applications to Ku band
- 0.15 micron GaAs PHEMT for applications to high-Ka band
- GaAs PIN process for low loss power switching applications
- Future new processes: InP HBT, SiGe, GaN, RF complementary metal oxide conductor (CMOS), RF MEMS

Figure 10.5 presents a GaAs wafer layout. Wafer size may be 3, 5, or 6 inches.

10.3.3   TYPES OF COMPONENTS DESIGNED

- Amplifiers—Low noise amplifiers (LNA), general, power amplifiers, wideband power amplifiers, distributed Traveling wave amplifiers (TWA)
- Mixers—balanced, star, sub-harmonic
- Switches—PIN, PHEMT, transmitting/receiving (T/R), matrix
- Frequency multipliers—active, passive
- Modulators—quadrature phase shift keying modulation (QPSK), quadrature amplitude modulation (QAM) (using PIN and PHEMT technology)
- Multifunction—receiving (RX) chip, transmitting (TX) chip, switched amplifiers chip, local oscillator (LO) chain

Table 10.1 presents types of devices fabricated by using MMIC technology.

TABLE 10.1
MMIC Technology

Material	FET	BJT	Diode
III–V-based	PHEMT GaAs	HBT GaAs	Schotky GaAs
	HEMT InP	D-HBT InP	
	MHEMT GaAs		
	HEMT GaN		
Silicon	CMOS	HBT SiGe	

Note: CMOS, complementary metal oxide semiconductor; D-HBT, double hetero-
structure bipolar transistor; HEMT, high electron mobility transistor;
MHEMT, metamorphic HEMT; PHEMT, pseudomorphic HEMT.

10.3.4 Advantages of GaAs versus Silicon

MMICs were originally fabricated by using gallium arsenide (GaAs), a III–V compound semiconductor. MMICs are dimensionally small (from around 1 mm^2 to
10 mm^2) and can be mass produced. GaAs has some electronic properties that are
better than those of silicon. It has a higher saturated electron velocity and higher
electron mobility, allowing transistors made from GaAs to function at frequencies
higher than 250 GHz. Unlike silicon junctions, GaAs devices are relatively insensi-
tive to heat because of their higher band gap. Also, GaAs devices tend to have less
noise than silicon devices, especially at high frequencies, which is a result of higher
carrier mobility and lower resistive device parasitic effects. These properties recom-
mend GaAs circuitry in mobile phones, satellite communications, microwave point-
to-point links, and higher frequency radar systems. It is used in the fabrication of
Gunn diodes to generate microwave. GaAs has a direct band gap, which means that
it can be used to emit light efficiently. Silicon has an indirect band gap and so is very
poor at emitting light. Nonetheless, recent advances may make silicon light-emitting
diodes (LEDs) and lasers possible. Owing to its lower band gap though, Si LEDs
cannot emit visible light and rather work in the infrared (IR) range while GaAs
LEDs function in visible red light. As a wide direct band gap material and resulting
resistance to radiation damage, GaAs is an excellent material for space electronics
and optical windows in high-power applications.

Silicon has three major advantages over GaAs for integrated circuit manufacturers. First, silicon is an inexpensive material. In addition, a Si crystal has an extremely
stable structure mechanically and it can be grown to very large diameter boules and
processed with very high yields. It is also a decent thermal conductor, thus enabling
very dense packing of transistors, all very attractive for design and manufacturing
of very large integrated circuits. The second major advantage of Si is the existence
of silicon dioxide—one of the best insulators. Silicon dioxide can easily be incorporated onto silicon circuits, and such layers are adherent to the underlying Si. GaAs
does not easily form such a stable adherent insulating layer and does not have a stable

TABLE 10.2
Comparison of Material Properties

Property	Si	Si or Sapphire	GaAs	InP
Dielectric constant	11.7	11.6	12.9	14
Resistivity (Ω/cm)	10^3–10^5	$>10^{14}$	10^7–10^9	10^7
Mobility (cm^2/V-s)	700	700	4300	3000
Density (gr/cm^3)	2.3	3.9	5.3	4.8
Saturation velocity (cm/s)	9×10^6	9×10^6	1.3×10^7	1.9×10^7

oxide either. The third, and perhaps most important, advantage of silicon is that it possesses a much higher hole mobility. This high mobility allows the fabrication of higher-speed P-channel field effect transistors, which are required for CMOS logic. Because they lack a fast CMOS structure, GaAs logic circuits have much higher power consumption, which has made them unable to compete with silicon logic circuits. The primary advantage of Si technology is its lower fabrication cost compared with GaAs. Silicon wafer diameters are larger, typically 8 or 12 inches compared with 4 or 6 inches for GaAs. Si wafer costs are much lower than GaAs wafer costs, contributing to a less expensive Si integrated circuit.

Other III–V technologies, such as indium phosphide (InP), offer better performance than GaAs in terms of gain, higher cutoff frequency, and low noise. However, they are more expensive because of smaller wafer sizes and increased material fragility.

Silicon germanium (SiGe) is an Si-based compound semiconductor technology offering higher speed transistors than conventional Si devices but with similar cost advantages.

Gallium nitride (GaN) is also an option for MMICs. Because GaN transistors can operate at much higher temperatures and work at much higher voltages than GaAs transistors, they make ideal power amplifiers at microwave frequencies. In Table 10.2 properties of materials used in MMIC technology are compared.

10.3.5 SEMICONDUCTOR TECHNOLOGY

Cutoff frequency of Si CMOS MMIC devices is lower than 200 GHz. Si CMOS MMIC devices are usually low-power and low-cost devices. The cutoff frequency of SiGe MMIC devices is lower than 200 GHz. SiGe MMIC devices are used as medium-power high-gain devices. Cutoff frequency of InP HBT devices is lower than 400 GHz. InP HBT devices are used as medium-power high-gain devices. The cutoff frequency of InP HEMT devices is lower than 600 GHz. InP HEMT devices are used as medium-power high-gain devices. In Table 10.3 properties of MMIC technologies are compared. Figure 10.6 presents a 0.15 micron PHEMT on a GaAs substrate.

10.3.6 MMIC FABRICATION PROCESS

The MMIC fabrication process consists of several controlled processes in a semiconductor fabrication plant. The process is listed in the next paragraph. In Figure 10.7

TABLE 10.3

Summary of Semiconductor Technology

	Si CMOS	SiGe HBT	InP HBT	InP HEMT	GaN HEMT
Cutoff frequency	>200 GHz	>200 GHz	>400 GHz	>600 GHz	>200 GHz
Published MMICs	170 GHz	245 GHz	325 GHz	670 GHz	200 GHz
Output power	Low	Medium	Medium	Medium	High
Gain	Low	High	High	Low	Low
RF noise	High	High	High	Low	Low
Yield	High	High	Medium	Low	Low
Mixed signal	Yes	Yes	Yes	No	No
1/f noise	High	Low	Low	High	High
Breakdown voltage	–1 V	–2 V	–4 V	–2 V	>20 V

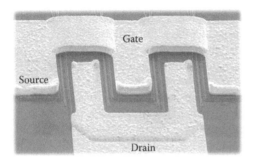

FIGURE 10.6   0.15 micron PHEMT on GaAs substrate.

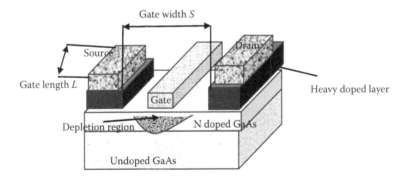

FIGURE 10.7   MESFET cross section on GaAs substrate.

a metal-semiconductor field effect transistor (MESFET) cross section on a GaAs substrate is shown.

MMIC Fabrication Process List
   Wafer fabrication: Preparing the wafer for fabrication.
   Wet cleans: Wafer cleaning by wet process.

Ion implantation: Dopants are embedded to create regions of increased or decreased conductivity. Selectively implant impurities. Create p- or n-type semiconductor regions.

Dry etching: Selectively remove materials.

Wet etching: Selectively remove materials by chemical process.

Plasma etching: Selectively remove materials.

Thermal treatment: High-temperature process to remove stress.

Rapid thermal anneal: High-temperature process to remove stress.

Fu-mace anneal: After ion implantation, thermal annealing is required. Furnace annealing may take minutes and causes too much diffusion of dopants for some applications.

Oxidation: Substrate oxidation, for example, dry oxidation $Si + O_2 \rightarrow SiO_2$; wet oxidation $Si + 2H_2O \rightarrow SiO_2 + 2H_2$.

Chemical vapor deposition (CVD): Chemical vapor deposited on the wafer. Pattern defined by photoresist.

Physical vapor deposition (PVD): Vapor produced by evaporation or sputtering deposited on the wafer. Pattern defined by photoresist.

Molecular beam epitaxy (MBE): A beam of atoms or molecules produced in high vacuum. Selectively grow layers of materials. Pattern defined by photoresist.

Electroplating: Electromechanical process used to add metal.

Chemical mechanical polish (CMP)

Wafer testing: Electrical test of the wafer.

Wafer back-grinding

Die preparation

Wafer mounting

Die cutting

Lithography: The process of transferring a pattern onto the wafer by selectively exposing and developing photoresist. Photolithography consists of four steps; the order depends on whether we are etching or lifting off the unwanted material.

Contact lithography: A glass plate is used that contains the pattern for the entire wafer. It is literally led against the wafer during exposure of the photoresist. In this case the entire wafer is patterned in one shot.

Electron-beam lithography: A form of direct-write lithography. Using E-beam lithography one can write directly to the wafer without a mask. Because an electron beam is used, rather than light, much smaller features can be resolved. Exposure can be done with light, UV light, or electron beam, depending on the accuracy needed. The E beam provides much higher resolution than light, because the particles are larger (greater momentum) and the wavelength is shorter.

Etching versus lift-off removal processes: There are two principal means of removing material, etching and lift-off.

The steps for an etch-off process are

1. Deposit material
2. Deposit photoresist
3. Pattern (expose and develop)
4. Remove material where it is not wanted by etching

Etching can be isotropic (etching wherever we can find the material we like to etch) or anisotropic (directional, etching only where the mask allows). Etches can be dry (reactive ion etching [RIE]) or wet (chemical). Etches can be very selective (etching only what we intend to etch) or nonselective (attacking a mask to the substrate).

In a lift-off process, the photoresist forms a mold, into which the desired material is deposited. The desired features are completed when photoresist B under unwanted areas is dissolved, and unwanted material is "lifted off."

Lift-off process steps are

- Deposit photoresist
- Pattern
- Deposit material conductor or insulator
- Remove material where it is not wanted by lifting off

In Figure 10.8 MESFET cross section on a GaAs substrate is shown. In Figure 10.9 a MMIC resistor cross section is shown. In Figure 10.10 an MMIC capacitor cross section is shown. Figure 10.11 presents the ion implantation process. Figure 10.12 presents the ion etch process. Figure 10.13 presents the wet etch process.

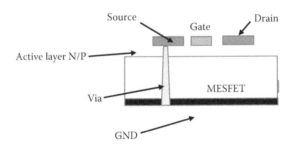

FIGURE 10.8   MESFET cross section.

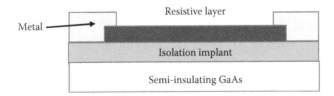

FIGURE 10.9   Resistor cross section.

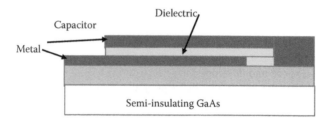

FIGURE 10.10   Capacitor cross section.

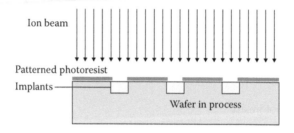

FIGURE 10.11   Ion implantation.

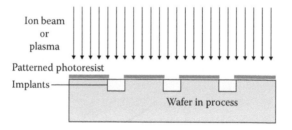

FIGURE 10.12   Ion etch.

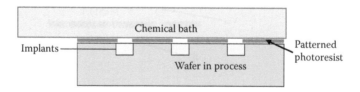

FIGURE 10.13   Wet etch.

10.3.7   GENERATION OF MICROWAVE SIGNALS IN MICROWAVE AND MM WAVE

Microwave signals can be generated by solid-state devices and vacuum tube based devices. Solid-state microwave devices are based on semiconductors such as silicon or GaAs, and include field-effect transistors (FETs), bipolar junction transistors (BJTs), Gunn diodes, and IMPATT diodes. Microwave variations of BJTs include the hetero-junction bipolar transistor (HBT), and microwave variants of FETs include the MESFET, the HEMT (also known as HFET), and LDMOS transistor. Microwaves can be generated and processed using integrated circuits, MMICs. They are usually manufactured using GaAs wafers, though SiGe and heavy-dope silicon are increasingly used. Vacuum tube based devices operate on the ballistic motion of electrons in a vacuum under the influence of controlling electric or magnetic fields, and include the magnetron, klystron, traveling wave tube (TWT), and gyrotron. These devices work in the density-modulated mode, rather than the current modulated mode. This means that they work on the basis of clumps of electrons flying ballistically through them, rather than using a continuous stream.

10.3.8 MMIC CIRCUIT EXAMPLES AND APPLICATIONS

Figure 10.14 presents a wide band mm wave power amplifier. The input power is divided by using a power divider. The RF signal is amplified by power amplifiers and combined by a power combiner to get the desired power at the device output. Figure 10.15 presents a wide band mm wave up converter. Figure 10.16 presents a Ka band PIN diode nonreflective single pole double through (SPDT). MMIC process costs are listed in Table 10.4.

MMIC Applications

• Ka band satellite communication
• 60-GHz wireless communication
• Automotive radars
• Imaging in security
• G-bit WLAN

10.4   MEMS TECHNOLOGY

Micro-electro-mechanical systems (MEMS) is the integration of mechanical elements, sensors, actuators, and electronics on a common silicon substrate through microfabrication technology. These devices replace bulky actuators and sensors with micron-scale equivalents that can produce large quantities by the fabrication process used in integrated circuits in photolithography. They reduce cost, bulk, weight, and

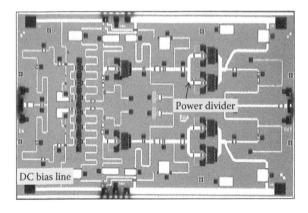

FIGURE 10.14   Wideband power amplifier.

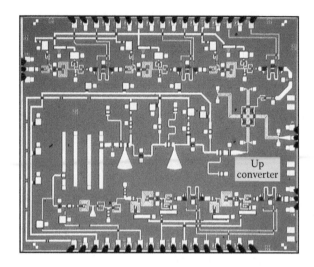

FIGURE 10.15   Ka band up converter.

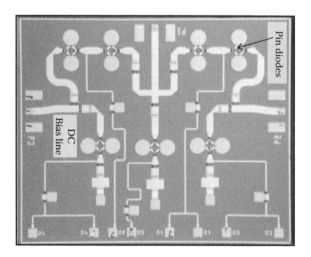

FIGURE 10.16   Ka band nonreflective SPDT.

TABLE 10.4
MMIC Cost

	Si CMOS	SiGe HBT	GaAs HEMT	InP HEMT
Chip cost ($/mm²)	0.01	0.1–0.5	1–2	10
Mask cost (M$/mask set)	1.35	0.135	0.0135	0.0135

power consumption while increasing performance, production volume, and functionality by orders of magnitude.

The electronics are fabricated using integrated circuit process sequences (e.g., CMOS, bipolar, or BICMOS processes); the micromechanical components are fabricated using compatible "micromachining" processes that selectively etch away parts of the silicon wafer or add new structural layers to form the mechanical and electromechanical devices.

10.4.1 MEMS Technology Advantages

- Low insertion loss <0.1 dB
- High isolation >50 dB
- Low distortion
- High linearity
- Very high Q
- Size reduction, system-on-a-chip
- High power handling ~40 dBm
- Low power consumption (~mW and no LNA)
- Low-cost high-volume fabrication

10.4.2 MEMS Technology Process

Bulk micromachining fabricates mechanical structures in the substrate by using orientation-dependent etching. Bulk micromachined substrate is presented in Figure 10.17. Surface micromachining fabricates mechanical structures above the substrate surface by using a sacrificial layer. A surface micro-machined substrate is presented in Figure 10.18.

In the bulk micromachining process silicon is machined using various etching processes. Surface micromachining uses layers deposited on the surface of a substrate as the structural materials, rather than using the substrate itself. The surface micromachining technique is relatively independent of the substrate used, and therefore can be easily mixed with other fabrication techniques that modify the substrate

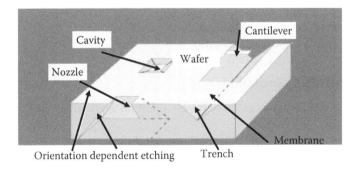

FIGURE 10.17 Bulk micromachining.

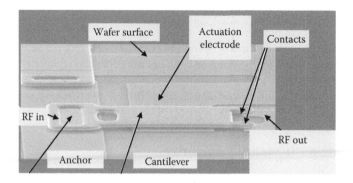

FIGURE 10.18   Surface micromachining.

first. An example is the fabrication of MEMS on a substrate with embedded control circuitry, in which MEMS technology is integrated with integrated circuit technology. This is being used to produce a wide variety of MEMS devices for many different applications. On the other hand, bulk micromachining is a subtractive fabrication technique, which converts the substrate, typically a single-crystal silicon, into the mechanical parts of the MEMS device. The MEMS device is first designed with a computer-aided design (CAD) tool. The design outcome is a layout and masks that are used to fabricate the MEMS device. The MEMS fabrication process is presented in Figure 10.19. MEMS fabrication technology is summarized in Table 10.5. In Figure 10.20 the block diagram of a MEMS bolometer coupled antenna array is presented.

Packaging of the device tends to be more difficult, but structures with increased heights are easier to fabricate when compared to surface micromachining. This is because the substrates can be thicker, resulting in relatively thick unsupported devices. Applications of RF MEMS technology are

- Tunable RF MEMS inductor
- Low loss switching matrix
- Tunable filters
- Bolometer coupled antenna array
- Low cost W-band detection array

10.4.3   MEMS COMPONENTS

MEMS components are categorized in one of several applications, such as

1. *Sensors* are a class of MEMS that are designed to sense changes and interact with their environments. These classes of MEMS include chemical, motion, inertia, thermal, RF sensors, and optical sensors. Microsensors are useful because of their small physical size, which allows them to be less invasive.

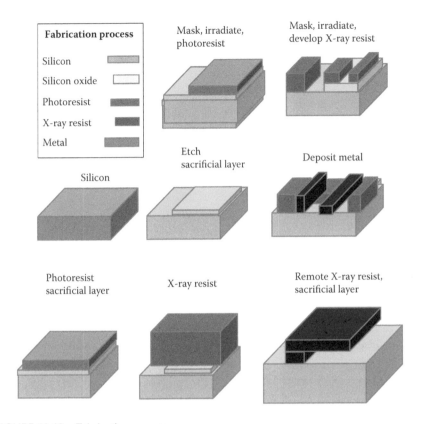

FIGURE 10.19   Fabrication process.

TABLE 10.5
Fabrication Technology

Technology	Process
Surface micromachining	Release and drying systems to realize free-standing microstructures.
Bulk micromachining	Dry etching systems to produce deep 2D free-form geometries with vertical sidewalls in substrates. Anisotropic wet etching systems with protection for wafer front sides during etching. Bonding and aligning systems to join wafers and perform photolithography on the stacked substrates.

2. *Actuators* are a group of devices designed to provide power or stimulus to other components or MEMS devices. MEMS actuators are either electrostatically or thermally driven.
3. *RF MEMS* are a class of devices used to switch or transmit high frequency, RF signals. Typical devices include metal contact switches, shunt switches, tunable capacitors, antennas, etc.
4. *Optical MEMS* are devices designed to direct, reflect, filter, and/or amplify light. These components include optical switches and reflectors.

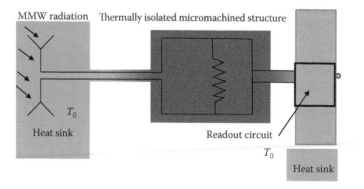

FIGURE 10.20 Bolometer coupled antenna array.

5. *Microfluidic MEMS* are devices designed to interact with fluid-based environments. Devices such as pumps and valves have been designed to move, eject, and mix small volumes of fluid.
6. *Bio MEMS* are devices that, much like microfluidic MEMS, are designed to interact specifically with biological samples. Devices such as these are designed to interact with proteins, biological cells, medical reagents, etc. and can be used for drug delivery or other in situ medical analyses.

10.5 LTCC AND HTCC TECHNOLOGY

Co-fired ceramic devices are monolithic, ceramic microelectronic devices in which the entire ceramic support structure and any conductive, resistive, and dielectric materials are fired in a kiln at the same time. Typical devices include capacitors, inductors, resistors, transformers, and hybrid circuits. The technology is also used for a multilayer packaging for the electronics industry, such as military electronics. Co-fired ceramic devices are made by processing a number of layers independently and assembling them into a device as a final step. Co-firing can be divided into low-temperature (LTCC) and high-temperature (HTCC) applications. Low temperature means that the sintering temperature is below 1000°C (1830°F), while high temperature is around 1600°C (2910°F). There are two types of raw ceramics to manufacture a multilayer ceramic (MLC) substrate:

- Ceramics fired at high temperature (T ≥ 1500°C): HTCC
- Ceramics fired at low temperature (T ≤ 1000°C): LTCC

The base material of HTCC is usually Al_2O_3. HTCC substrates are row ceramic sheets. Because of the high firing temperature of Al_2O_3 the material of the embedded layers can only be high melting temperature metals: wolfram, molybdenum, or manganese. The substrate is unsuitable to bury passive elements, although it is possible to produce thick-film networks and circuits on the surface of an HTCC ceramic.

The breakthrough for LTCC fabrication was when the firing temperature of the ceramic–glass substrate was reduced to 850°C. The equipment for conventional thick-film process could be used to fabricate LTCC devices. LTCC technology

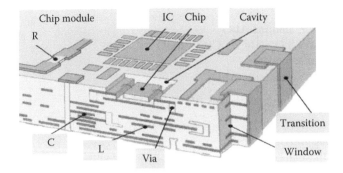

FIGURE 10.21   LTCC module.

TABLE 10.6
Dielectric Properties at 9 GHz
of LTCC Substrates

Material	ε_r	Tan $\delta \times 10^{-3}$
99.5% Al	9.98	0.1
LTCC1	7.33	3.0
LTCC2	6.27	0.4
LTCC3	7.2	0.6
LTCC4	7.44	1.2
LTCC5	6.84	1.3
LTCC6	8.89	1.4

evolved from HTCC technology and combined the advantageous features of thick-film technology. Because of the low firing temperature (850°C) the same materials are used for producing buried and surface wiring and resistive layers as thick-film hybrid integrated circuits (i.e., Au, Ag, Cu wiring RuO_2-based resistive layers). It can be fired in an oxygen-rich environment, unlike HTCC boards, in which a reduced atmosphere is used. During co-firing the glass melts, and the conductive and ceramic particles are sintered. On the surface of LTCC substrates hybrid integrated circuits can be realized, shown in Figure 10.21. Passive elements can be buried into the substrate, and we can place semiconductor chips in a cavity. Dielectric Properties at 9GHz of LTCC substrates are listed in Table 10.6.

10.5.1   LTCC AND HTCC TECHNOLOGY PROCESS

- Low temperature (LTCC 875°C)
- High temperature (HTCC 1400°C–1600°C)
- Co-fired
   - Co-firing of (di)electric pastes
   - LTCC: precious metals (Au, Ag, Pd, Cu)
   - HTCC: refractory metals (W, Mo, MoMn)

- Ceramic
  Mix of: alumina Al_2O_3
  Glasses $SiO_2 - B_2O_3 - CaO - MgO$
  Organic binders
  HTTC: essentially Al_2O_3

Advantages of LTCC:

- Low permittivity tolerance
- Good thermal conductivity
- Low temperature co-firing environment (adapted to silicon and GaAs)
- Excellently suited for multilayer modules
- Integration of cavities and passive elements such as R, L, and C components
- Very robust against mechanical and thermal stress (hermetically sealed)
- Compose-able with fluidic, chemical, thermal, and mechanical functionalities
- Low material costs for silver conductor paths
- Low production costs for medium and large quantities

LTTC Advantages for high-frequency applications:

- Parallel processing (high yield, fast turnaround, lower cost)
- Precisely defined parameters
- High-performance conductors
- Potential for multilayer structures
- High interconnect density

LTCC process steps are listed as follows. LTCC raw material comes as sheets or rolls. Material manufacturers are DuPont, ESL, Ferro, and Heraeus.

- Tape casting
- Sheet cutting
- Laser punching
- Printing
- Cavity punching
- Stacking
- Bottom side printing
- Pressing
- Side hole formation
- Side hole printing
- Snap line formation
- Pallet firing
- Plating Ni-Au

In Figure 10.22 an LTCC process block diagram is presented. In Table 10.7 several electrical, thermal, and mechanical characteristics of several LTCC materials are listed.

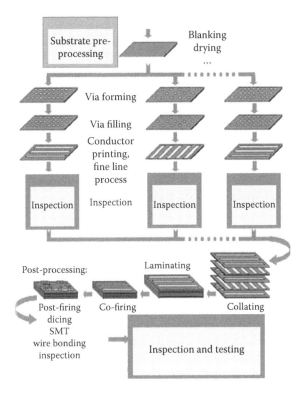

FIGURE 10.22 LTCC process.

TABLE 10.7
LTCC Material Characteristics

Material	LTCC DP951	Al$_2$O$_3$ 96%	BeO	AlN 98%
Electrical Characteristics at 10 MHz				
Dielectric constant, ε_r	7.8	9.6	6.5	8.6
Dissipation factor, tan δ	0.00015	0.0003	0.0002	0.0005
Thermal Characteristics				
Thermal expansion 10^{-6}/°C	5.8	7.1	7.5	4.6
Thermal conductivity W/mk 25°C–300°C	3	20.9	251	180
Mechanical Characteristics				
Density	3.1	3.8	2.8	3.3
Flexural strength (MPa)	320	274	241	340
Young's modulus (GPa)	120	314	343	340

TABLE 10.8
LTCC Line Loss

Material	Dissipation Factor, tan δ 10⁻³	Line Loss (dB/mm), at 2 GHz
LTTC1	3.8	0.004
LTTC2	2.0	0.0035
LTTC6-CT2000	1.7	0.0033
Alumina 99.5%	0.65	0.003

In Table 10.8 LTCC line losses at 2 GHz are listed for several LTCC materials. For LTTC1 material losses are 0.004 dB/mm.

10.5.2 DESIGN OF HIGH-PASS LTCC FILTERS

The trend in the wireless industry toward miniaturization, cost reduction, and improved performance drives the microwave designer to develop microwave components in LTCC technology. A significant reduction in the size and the cost of microwave components may be achieved by using LTCC technology. In LTCC technology discrete surface-mounted components such as capacitors and inductors are replaced by integrated printed components. LTCC technology allows the designer to use a multilayer design if needed to reduce the size and cost of the circuit. However, a multilayer design results in more losses due to via connections and due to parasitic coupling between different parts of the circuit. To improve the filter performance all the filter parameters have been optimized. Package effects were taken into account in the design.

High Pass Filter Specification
    Frequency—1.5–2.5 GHz
    Insertion loss at 1.1 Fo—1 dB
    Rejection at 0.9 Fo—3 dB
    Rejection at 0.75 Fo—20 dB
    Rejection at 0.5 Fo—40 dB
    VSWR—2:1
    Case dimensions—700 × 300 × 25.5 mil-inch

The filters are realized by using lumped elements. The filter inductors and capacitor parameters were optimized by using HP ADS software. The filter consists of five layers of a 5.1 mil substrate with $\varepsilon_r = 7.8$. Package effects were taken into account in the design. Changes in the design were made to compensate and minimize package effects. In Figure 10.23 the filter layout is presented. S_{11} and S_{12} momentum simulation results are shown in Figure 10.24. In Figure 10.25 the filter 2 layout is presented. S_{11} and S_{12} momentum simulation results are shown in Figure 10.26. Simulation results of a tolerance check are shown in Figure 10.27. The parameters that were tested in the tolerance check are inductor and capacitor line width and length and spacing between capacitor fingers.

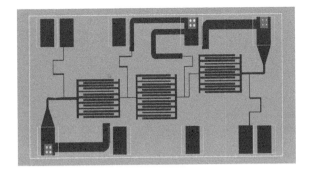

FIGURE 10.23 Layout of high-pass filter no. 1.

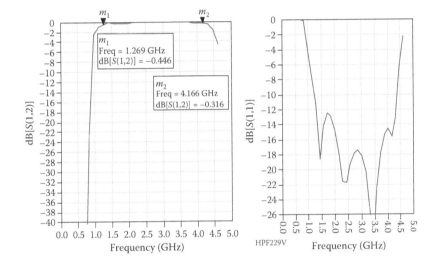

FIGURE 10.24 S_{12} and S_{11} results of high-pass filter no. 1.

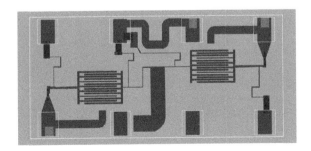

FIGURE 10.25 Layout of high-pass filter no. 2.

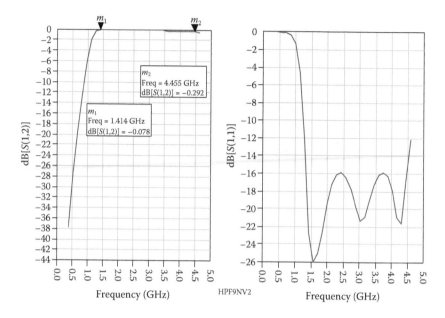

FIGURE 10.26 S_{12} and S_{11} results of high-pass filter no. 2.

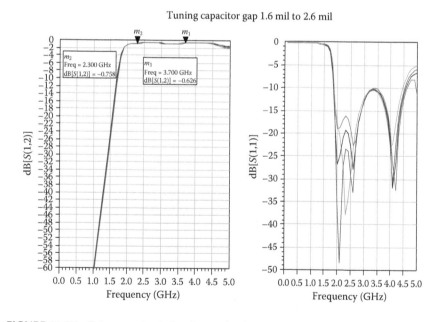

FIGURE 10.27   Tolerance simulation for spacing between capacitor fingers.

10.5.3 COMPARISON OF SINGLE-LAYER AND MULTILAYER MICROSTRIP CIRCUITS

In a single-layer microstrip circuit all conductors are in a single layer. Coupling between conductors is achieved through edge or end proximity (across narrow gaps). Single-layer microstrip circuits are inexpensive in production. In Figure 10.28 a single-layer microstrip edge coupled filter is shown. Figure 10.29 presents the layout of a single-layer microstrip directional coupler. Figure 10.30 presents the structure of a multilayer microstrip coupler.

In multilayer microwave circuits conductors are separated by dielectric layers and stacked on different layers. This structure allows for (strong) broadside coupling. Registration between layers is not difficult to achieve as narrow gaps between strips in single layer circuits. The multilayer structure technique is well suited to thick-film print technology and also suitable for LTCC technology.

FIGURE 10.28   Edge coupled filter.

FIGURE 10.29   Single-layer microstrip directional coupler.

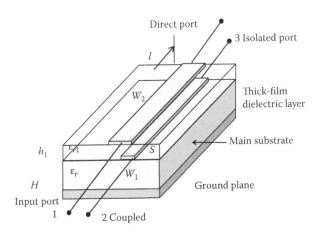

FIGURE 10.30   Multilayer microstrip coupler.

10.6　CONCLUSIONS

Dimensions and losses of microwave systems are minimized by using MIC, MMIC, MEMS, and LTCC technology. Dimensions and losses of microwave systems are minimized by using the multilayer structure technique. The multilayer structure technique is well suited to thick-film print technology and for LTCC technology. LTCC technology allows integration of cavities and passive elements such as R, L, and C components as part of the LTCC circuits. Sensors, actuators, and RF switches may be manufactured by using MEMS technology. Losses of MEMS components are considerably lower than those using MIC and MMIC RF components. MMICs are circuits in which active and passive elements are formed on the same dielectric substrate. MMICs are dimensionally small (from around 1 mm^2 to 10 mm^2) and can be mass produced. MMIC components cannot be tuned. An accurate design is crucial in the design of MMIC circuits. The goal of MMIC, MEMS, and LTCC designers is to comply with customer specifications in one design iteration.

REFERENCES

1. Rogers, J., and Plett, C. *Radio frequency Integrated Circuit Design.* Norwood, MA: Artech House, 2003.
2. Maluf, N., and Williams, K. *An Introduction to Microelectromechanical System Engineering.* Norwood, MA: Artech House, 2004.
3. Sabban, A. Microstrip Antenna Arrays. In Nasimuddin Nasimuddin (Ed.), *Microstrip Antennas*, pp. 361–384, 2011. InTech, http://www.intechopen.com/articles/show/title/microstrip-antenna-arrays.
4. Sabban, A. Applications of MM Wave Microstrip Antenna Arrays. In *ISSSE 2007 Conference*, Montreal, August 2007.
5. Gauthier, G. P., Raskin, G. P., Rebiez, G. M., and Kathei, P. B. A 94 GHz Micro-machined Aperture-Coupled Microstrip Antenna. *IEEE Transactions on Antenna and Propagation*, 47(12):1761–1766, 1999.
6. Milkov, M. M. *Millimeter-Wave Imaging System Based on Antenna-Coupled Bolometer.* MSc. thesis, UCLA, 2000.
7. de Lange, G. et al., A 3*3 mm-Wave Micro Machined Imaging Array with Sis Mixers. *Applied Physics Letters,* 75(6):868–870, 1999.
8. Rahman, A. et al. Micro-machined Room Temperature Micro Bolometers for MM-Wave Detection. *Applied Physics Letters*, 68(14):2020–2022, 1996.
9. Mass, S. A. *Nonlinear Microwave and RF Circuits.* Norwood, MA: Artech House, 1997.

11 Radio Frequency Measurements

11.1 INTRODUCTION

S-parameter measurements are the first stage in electromagnetics and antenna measurements. It is more convenient to measure antenna measurements in the receiving mode. If the measured antenna is reciprocal the antenna radiation characteristics are identical for the receiving and transmitting modes. Active antennas are not reciprocal. Radiation characteristics of antennas are usually measured in the far field. Far-field antenna measurements suffer from some disadvantages. A long free space area is needed. Reflections from the ground from walls affect measured results and add errors to measured results. It is difficult and almost impossible to measure the antenna on the antenna operating environment, such as an airplane or satellite. Antenna measurement facilities are expensive. Some of these drawbacks may be solved by near-field and indoor measurements. Near-field measurements are presented in Ref. [1]. Small communication companies do not own antenna measurement facilities. However, there are several companies around the world that provide antenna measurement services, near-field and far-field measurements. One day near-field measurements may cost around US$5000. One day far-field measurements may cost around US$2000.

11.2 MULTIPORT NETWORKS WITH N PORTS

Antenna systems and communication systems may be represented as multiport networks with N ports as shown in Figure 11.1. We may assume that only one mode propagates in each port. The electromagnetic fields in each port represent incident and reflected waves. The electromagnetic fields may be represented by equivalent voltages and currents as given in Equations 11.1 and 11.2.

$$V_n^- = Z_n I_n^-$$
$$V_n^+ = Z_n I_n^+ \tag{11.1}$$

$$I_n^- = Y_n V_n^-$$
$$I_n^+ = Y_n V_n^+ \tag{11.2}$$

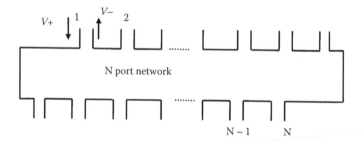

FIGURE 11.1 Multiport networks with N ports.

The voltages and currents in each port are given in Equation 11.3.

$$I_n = I_n^+ - I_n^-$$
$$V_n = V_n^+ + V_n^-$$

(11.3)

The relations between the voltages and currents may be represented by the Z matrix as given in Equation 11.4. The relations between the currents and voltages may be represented by the Y matrix as given in Equation 11.5. The Y matrix is the inverse of the Z matrix.

$$[V] = \begin{bmatrix} V_1 \\ V_2 \\ V_N \end{bmatrix} = \begin{bmatrix} Z_{11} & Z_{12} & Z_{1N} \\ Z_{21} & Z_{22} & Z_{2N} \\ Z_{N1} & Z_{N2} & Z_{NN} \end{bmatrix} \begin{bmatrix} I_1 \\ I_2 \\ I_N \end{bmatrix} = [Z][I]$$

(11.4)

$$[I] = \begin{bmatrix} I_1 \\ I_2 \\ I_N \end{bmatrix} = \begin{bmatrix} Y_{11} & Y_{12} & Y_{1N} \\ Y_{21} & Y_{22} & Y_{2N} \\ Y_{N1} & Y_{N2} & Y_{NN} \end{bmatrix} \begin{bmatrix} V_1 \\ V_2 \\ V_N \end{bmatrix} = [Y][V]$$

(11.5)

11.3 SCATTERING MATRIX

We cannot measure voltages and currents in microwave networks. However, we can measure power, voltage standing wave ratio (VSWR), and the location of the minimum field strength. We can calculate the reflection coefficient from these data. The scattering matrix is a mathematical presentation that describes how electromagnetic energy propagates through a multiport network. The S matrix allows us to accurately describe the properties of complicated networks. S parameters are defined for a given frequency and system impedance, and vary as a function of frequency for any nonideal network. The scattering S matrix describes the relation between the forward and reflected waves as written in Equation 11.6. S parameters describe the response of an N-port network to voltage signals at each port. The first number in the subscript refers to the responding port, while the second number refers to the incident port. Thus S_{21} means the response at port 2 due to a signal at port 1.

$$[V^-] = \begin{bmatrix} V_1^- \\ V_2^- \\ V_2^+ \end{bmatrix} = \begin{bmatrix} S_{11} & S_{12} & S_{1N} \\ S_{21} & S_{22} & S_{2N} \\ S_{N1} & S_{N2} & S_{NN} \end{bmatrix} \begin{bmatrix} V_1^+ \\ V_2^+ \\ V_2^+ \end{bmatrix} = [S][V^+] \tag{11.6}$$

The S_{nn} elements represent reflection coefficients. The S_{nm} elements represent transmission coefficients as written in Equation 11.7, where a_i represents the forward voltage in the i port.

$$S_{nn} = \frac{V_n^-}{V_n^+}\Big|a_i = 0 \quad i \neq n$$

$$S_{nm} = \frac{V_n^-}{V_m^+}\Big|a_i = 0 \quad i \neq m \tag{11.7}$$

By normalizing the S matrix we can represent the forward and reflected voltages as written in Equation 11.8. S parameters depend on the frequency and are given as function of frequency. In a reciprocal microwave network $S_{nm} = S_{mn}$ and $[S]' = [S]$.

$$I = I^+ - I^- = V^+ - V^-$$
$$V = V^+ + V^-$$
$$V^+ = \frac{1}{2}(V + I) \tag{11.8}$$
$$V^- = \frac{1}{2}(V - I)$$

The relation between the Z and the S matrix is derived by using Equations 11.8 and 11.9 and is given in Equations 11.10 and 11.11.

$$I_n = I_n^+ - I_n^- = V_n^+ - V_n^-$$
$$V_n = V_n^+ + V_n^- \tag{11.9}$$

$$[V] = [V^+] + [V^-] = [Z][I] = [Z][V^+] - [Z][V^-]$$
$$([Z] + [U])[V^-] = ([Z] - [U])[V^+]$$
$$[V^-] = ([Z] + [U])^{-1}([Z] - [U])[V^+] \tag{11.10}$$
$$[V^-] = [S][V^+]$$
$$[S] = ([Z] + [U])^{-1}([Z] - [U])$$

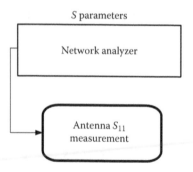

FIGURE 11.2   Antenna S parameter measurements.

$$[V^+] = \frac{1}{2}([V]+[I]) = \frac{1}{2}([Z]+[U])[I]$$

$$[V^-] = \frac{1}{2}([V]-[I]) = \frac{1}{2}([Z]-[U])[I]$$

$$\frac{1}{2}[I] = ([Z]+[U])^{-1}[V^+]$$ (11.11)

$$[V^-] = ([Z]-[U])([Z]+[U])^{-1}[V^+]$$

$$[S] = ([Z]-[U])([Z]+[U])^{-1}$$

A network analyzer is employed to measure S parameters, as shown in Figure 11.2. A network analyzer may have 2–16 ports.

11.4   S-PARAMETER MEASUREMENTS

Antenna S-parameter measurement is usually a one-port measurement. First we calibrate the network analyzer to the desired frequency range. A one-port, S_{1P}, calibration process consists of three steps.

- Short calibration
- Open calibration
- Load calibration

Connect the antenna to the network analyzer and measure the S_{11} parameter.

Save and plot S_{11} results. A setup for S-parameter measurement is shown in Figure 11.2.

A two-port S-parameter measurement setup is shown in Figure 11.3. A two-port, S_{2P}, calibration process consists of four steps.

- Short calibration
- Open calibration
- Load calibration
- Through calibration

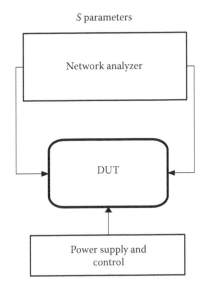

S parameters

Network analyzer

DUT

Power supply and control

FIGURE 11.3 Two-port S parameter measurements.

TABLE 11.1
S Parameter Results

Channel	S_{11} (E/T) dB	S_{22} (E/T) dB	S_{12} (E/T) dB	S_{21} (E/T) dB
1	−10	−10	−20	30
2	−10.5	−11	−21	29
3	−11	−10	−20	29
$n-1$	−10	−9	−20	29
n	−9	−10	−19	30

Measure the S parameters: S_{11}, S_{22}, S_{12}, and S_{21} for N channels. Table 11.1 presents a typical table of measured S parameter results. S parameters in decibels may be calculated by using Equation 11.12.

$$Sij(dB) = 20*log[Sij(magnitude)] \tag{11.12}$$

11.4.1 Types of S-Parameter Measurements

- **Small-signal S-parameter measurements:** In small signal S-parameter measurements the signals have only linear effects on the network so that gain compression does not take place. Passive networks are linear at any power level.
- **Large-signal S-parameter measurements:** S parameters are measured for different power levels. The S-matrix will vary with input signal strength.

11.5 TRANSMISSION MEASUREMENTS

A block diagram of transmission measurements setup is shown in Figure 11.4. The transmission measurement setup consists of a sweep generator, Device under Test, transmitting and receiving antennas, and a spectrum analyzer.

The received power may be calculated by using the Friis equation as given in Equations 11.13 and 11.14. The receiving antenna may be a standard gain antenna with a known gain, where r represents the distance between the antennas.

$$P_R = P_T G_T G_R \left(\frac{\lambda}{4\pi r} \right)^2$$

$$\text{For} - G_T = G_R = G \qquad (11.13)$$

$$G = \sqrt{\frac{P_R}{P_T}} \left(\frac{4\pi r}{\lambda} \right)$$

$$P_R = P_T G_T G_R \left(\frac{\lambda}{4\pi r} \right)^2$$

$$\text{For} - G_T \neq G_R \qquad (11.14)$$

$$G_T = \frac{1}{G_R} \frac{P_R}{P_T} \left(\frac{4\pi r}{\lambda} \right)^2$$

Transmission measurement results may be summarized as in Table 11.2.

Transmitting channel measurements

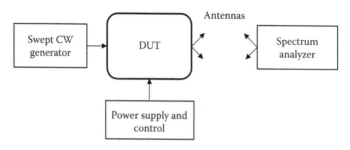

FIGURE 11.4 Transmission measurement setup.

TABLE 11.2
Transmission Measurement Results

Transmission Results for Antennas Under Test (AUT) dBm

Antenna	F1 (MHz)	F2 (MHz)	F3 (MHz)	Remarks
1	10	9	8	
2	9	8	7	
3	9.5	8.5	7.5	
4	10	9	8	
5	9	8	7	
6	10.5	9.5	8.5	
7	9	8	7	
8	11	10	9	

11.6 OUTPUT POWER AND LINEARITY MEASUREMENTS

A block diagram of output power and linearity measurement setup is shown in Figure 11.5. The output power and linearity measurement setup consists of a sweep generator, Device under Test, and a spectrum analyzer. In output power and linearity measurements we increase the synthesizer power in 1-dB steps and measure the output power level and linearity.

11.7 ANTENNA MEASUREMENTS

Typical parameters of antennas are radiation pattern, gain, directivity, beam width, polarization, and impedance. During antenna measurements we ensure that the antennas meet the required specifications and we can characterize the antenna parameters.

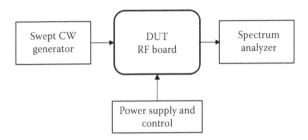

FIGURE 11.5 Output power and linearity measurement setup.

11.7.1 RADIATION PATTERN MEASUREMENTS

A radiation pattern is the antenna radiated field as function of the direction in space. The radiated field is measured at various angles at a constant distance from the antenna. The radiation pattern of an antenna can be defined as the locus of all points where the emitted power per unit surface is the same. The radiated power per unit surface is proportional to the square of the electric field of the electromagnetic wave. The radiation pattern is the locus of points with the same electrical field strength. Usually the antenna radiation pattern is measured in a far-field antenna range. The antenna under test is placed in the far-field distance from the transmitting antenna. Because of the size required to create a far-field range for large antennas near-field techniques are employed. Near-field techniques allow one to measure the fields on a surface close to the antenna (usually 3–10 wavelength). Near-field distances are transferred to far-field by using the Fourier transform.

The far-field distance or Fraunhofer distance, R, is given in Equation 11.15.

$$R = 2D^2/\lambda \qquad (11.15)$$

where D is the maximum antenna dimension and λ is the antenna wavelength.

The radiation pattern graphs can be drawn using Cartesian (rectangular) coordinates as shown in Figure 11.6 (see Refs. [1–4]). A polarplot is shown in Figure 11.7. The polarplot is useful to measure the beam width, which is the angle at the −3dB points around the maximum gain. A 3D radiation pattern is shown in Figure 11.8.

The main beam is the region around the direction of maximum radiation, usually the region that is within 3 dB of the peak of the main lobe.

The beam width is the angular range of the antenna pattern in which at least half of the maximum power is emitted. This angular range, of the major lobe, is defined as the points at which the field strength falls around 3 dB with reference to the maximum field strength.

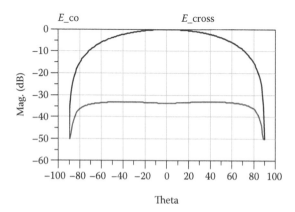

FIGURE 11.6 Radiation pattern of a loop antenna with ground plane rectangular plot.

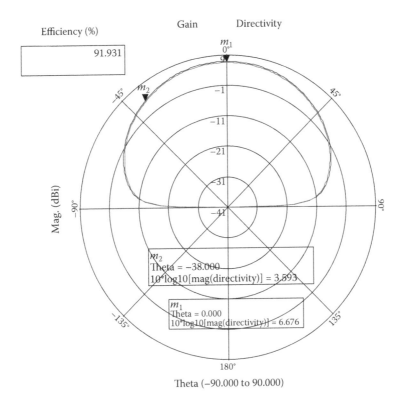

FIGURE 11.7   Grounded quarter wavelength patch antenna polar radiation pattern.

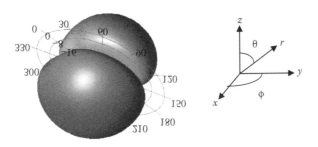

FIGURE 11.8   Loop antenna 3D radiation pattern.

Side lobes are smaller beams that are away from the main beam. Side lobes present radiation in undesired directions. The side-lobe level is a parameter used to characterize the antenna radiation pattern. It is the maximum value of the side lobes away from the main beam and is expressed usually in decibels.

The radiated power is the total radiated power when the antenna is excited by a current or voltage of known intensity.

11.7.2 Directivity and Antenna Effective Area

The ratio between the amounts of energy propagating in a certain direction compared to the average energy radiated to all directions over a sphere (see Refs. [1–4]) is

$$D = \frac{P(\theta, \phi)\text{maximal}}{P(\theta, \phi)\text{average}} = 4\pi \frac{P(\theta, \phi)\text{maximal}}{P \text{ rad}} \tag{11.16}$$

$$P(\theta, \phi)\text{average} = \frac{1}{4\pi} \iint P(\theta, \phi) \sin\theta \, d\theta \, d\phi = \frac{P \text{ rad}}{4\pi} \tag{11.17}$$

$$D \sim \frac{4\pi}{\theta E \times \theta H} \tag{11.18}$$

where
θE = measured beam width in radian in *EL* plane
θH = measured beam width in radian in *AZ* plane

Measured beam width in radian in *AZ* plane and in *EL* plane allow us to calculate antenna directivity. The antenna effective area (A_{eff}) is the antenna area which contributes to the antenna directivity.

$$A_{\text{eff}} = \frac{D\lambda^2}{4\pi} \sim \frac{\lambda^2}{\theta E \times \theta H} \tag{11.19}$$

11.7.3 Radiation Efficiency

Radiation efficiency is the ratio of power radiated to the total input power. The efficiency of an antenna takes into account losses, and is equal to the total radiated power divided by the radiated power of an ideal lossless antenna.

$$G = \alpha D \tag{11.20}$$

Efficiency is equal to the radiation resistance divided by total resistance (real part) of the feed-point impedance. Efficiency is defined as the ratio of the power that is radiated to the total power used by the antenna as given in Equation 11.21. Total power is equal to power radiated plus power loss.

$$\alpha = \frac{P_r}{P_r + P_l} \tag{11.21}$$

An *E* and *H* plane 3D radiation pattern of a wire loop antenna in free space is shown in Figure 11.9.

FIGURE 11.9 *E* and *H* plane radiation pattern of a loop antenna in free space.

11.7.4 TYPICAL ANTENNA RADIATION PATTERN

A typical antenna radiation pattern is shown in Figure 11.10. The antenna main beam is measured between the points that the maximum relative field intensity *E* decays to 0.707*E*. Half of the radiated power is concentrated in the antenna main beam. The antenna main beam is called the 3 dB beam width. Radiation to the undesired direction is concentrated in the antenna side lobes. The antenna radiation pattern is usually measured in free space ranges. An elevated free space range is shown in Figure 11.11. An anechoic chamber is shown in Figure 11.12.

11.7.5 GAIN MEASUREMENTS

The ratio between the amount of energy propagating in a certain direction compared to the energy that would be propagating in the same direction if the antenna were not directional, isotropic radiator, is known as its gain (G).

Figure 11.11 presents the antenna far-field range for radiation pattern measurements. Antenna gain is measured by comparing the field strength measured by the antenna under test to the field strength measured by a standard gain horn as shown

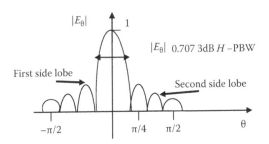

FIGURE 11.10 Typical antenna radiation pattern.

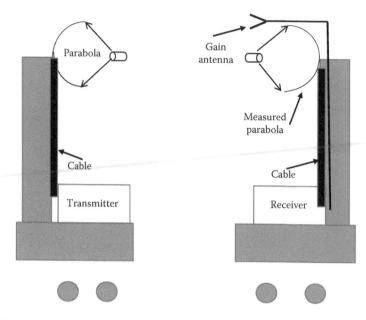

FIGURE 11.11   Antenna range for radiation pattern measurements.

FIGURE 11.12   Anechoic chamber.

in Figure 11.11. The gain as a function of frequency of the standard gain horn is supplied by the standard gain horn manufacturer. Figure 11.12 presents an anechoic chamber used to indoor antenna measurements. The chamber metallic walls are covered with absorbing materials.

11.8 ANTENNA RANGE SETUP

An antenna range setup is shown in Figure 11.11. The antenna range setup consists of the following instruments:

- Transmitting system that consists of a wideband signal generator and transmitting antenna
- Measured receiving antenna
- Receiver
- Positioning system
- Recorder and plotter
- Computer and data processing system

The signal generator should be stable with controlled frequency value, good spectral purity, and controlled power level. A low-cost receiving system consists of a detector and amplifiers. Several companies sell antenna measurement setups, for example, Agilent, Tektronix, Antritsu, and others.

REFERENCES

1. Balanis, C. A. *Antenna Theory: Analysis and Design.* 2nd ed. New York: John Wiley & Sons, 1996.
2. Godara, L. C. (ed.). *Handbook of Antennas in Wireless Communications.* Boca Raton, FL: CRC Press, 2002.
3. Kraus, J. D., and Marhefka, R. J. *Antennas for All Applications,* 3rd ed. New York: McGraw-Hill, 2002.
4. Sabban, A. *RF Engineering, Microwave and Antennas.* Tel Aviv, Israel: Saar Publication, 2014.

Index